JON TURNEY

This edition published in the UK in 2016 by
Icon Books Ltd, Omnibus Business Centre,
39–41 North Road, London N7 9DP
email: info@iconbooks.com
www.iconbooks.com

Originally published in the UK in 2015 by Icon Books Ltd

Sold in the UK, Europe and Asia
by Faber & Faber Ltd, Bloomsbury House,
74–77 Great Russell Street,
London WC1B 3DA or their agents

Distributed in the UK, Europe and Asia
by TBS Ltd, TBS Distribution Centre, Colchester Road,
Frating Green, Colchester CO7 7DW

Distributed in Australia and New Zealand
by Allen & Unwin Pty Ltd,
PO Box 8500, 83 Alexander Street,
Crows Nest, NSW 2065

Distributed in South Africa by
Jonathan Ball, Office B4, The District,
41 Sir Lowry Road, Woodstock 7925

Distributed in India by Penguin Books India,
7th Floor, Infinity Tower – C, DLF Cyber City,
Gurgaon 122002, Haryana

Distributed in the USA by
Publishers Group West,
1700 Fourth Street, Berkeley, CA 94710

Distributed in Canada by Publishers Group Canada,
76 Stafford Street, Unit 300
Toronto, Ontario M6J 2S1

ISBN: 978-178578-024-0

Typeset in Garamond by Marie Doherty
Printed and bound in the UK by Clays Ltd, St Ives plc

Contents

'with increasing knowledge,
revulsion sometimes yields to fascination'
—Theodor Rosebury, *Life on Man*

Introduction
Organism, meet superorganism

W ho am I? The mirror reminds me. Here I stand, unclothed (don't worry, I'll look so you don't have to). I see a pale, upright biped. White, middle-aged, male. Tall for a human, with a slight tendency to stoop. I've had a medically charmed life so far, as baby-boomers in the West may enjoy. Not too many signs of wear and tear yet. If I recall the image in the glass from 30 years ago, say, the current version of me has a thickening waist – and, I notice, toenails – and thinning hair. Otherwise I seem to look much the same.

That continuity, this body, is part of my sense of self. A human body, pretty large by the standards of Earth creatures, is a marker of individuality. I do not mean that in any philosophical sense – I am an autonomous human subject and this is my body. Or perhaps I do. I do not mean it, anyway, in any sense of the individual 'me' as unique.

But I do feel, for what it's worth, that the infinitesimally tiny fraction of the universe's matter and energy that is me can be *separated out*. The portion of biomass that I haul around – or that hauls me around – has the usual apertures and orifices where stuff goes in or comes out. If I want to stay alive, that has to happen with some regularity. But it looks to me as if this body has a pretty clear boundary. It seems clearly distinct from the rest of the world. I do experience conscious connections with other people, mainly because I have language. Today I am connected to thousands more, through our technologies. But the 'I' that is connected is also embodied. I am a biological entity; a human animal; an organism.

I have always felt that this organismic me, now one of 7 billion or so of the type, is a vessel built for a solo voyage. Artists and poets share that intuition. As Orson Welles said, we are born alone, we live alone, we die alone. He went on to say that love and friendship can still make life worthwhile – which is true. But the premise, it turns out, is entirely wrong. Science, quietly at first but with increasing insistence over the last decade or so, begs to differ. You, like me, are an individual human. But we've all got company. Lots of company.

Look again

Ever looked inside your mouth – not with a mirror, but a microscope? Chances are you did this at school. If you don't remember, here's what you do. Gently scrape the inside of your cheek. Use a clean cotton swab if you have one, the flat end of a toothpick or even the end of your fingernail if you don't. Dab the goo on a microscope slide, add a drop of methylene blue dye, top with a cover slip, slide onto the microscope stand and ... focus.

Even at low power, say ten times magnification, you'll now see some cells. They are flat – and not just because they are squashed on the slide. Flat is normal for cells of this type, from the tissue known as squamous epithelium. You can see the nucleus, which looks after the hereditary instructions (the dye stains the DNA, among other things), separate from the rest, the cytoplasm. A single cell at this magnification doesn't look that impressive. But every one is a reminder of an astonishing fact. Each of us is a vast, highly organised coalition of many such cells, tiny morsels of a much bigger organism, which can usually grow and divide in their own right. Far beneath the everyday scale of an upright mammal are the elementary particles of life. Growing and dividing from a single cell, a newly fertilised egg, they eventually number in the trillions,* and play their part in maintaining the highly organised assembly that is me, or you.

But there's more. Up the magnification, and there will be small blue-dyed dots under and around the large, blobbier-looking epithelial cells. They are bacteria. They will be there even if you used the clean swab rather than your finger. Your mouth – cheeks, tongue, teeth, gums and all – is full of them. It is warm and moist in there, and you keep adding nutrients, on their way to the stomach. What's not to like? You would find them, too, on every other surface of your body, including the internal ones like the intestine.

Until very recently, few gave much thought to the bacteria and other microbes to whom we give house room. But they have been there all through our evolution. Bacteria got here first. They are woven into our lives more closely than we ever imagined until recently. They congregate in complex, shifting communities that are shaped by, and help shape, the lives of our other cells. They carry out a surprisingly large portion of our

---

* Humans are squishy, and quite variable. So are cells. Result: it is pretty hard to count how many cells the average adult has. You'll find 10 trillion (10,000,000,000,000) to 100 trillion cited quite often as a rough total. A careful recent estimate, taking into account the volumes and main cell types of different organs, comes out at a bit over 37 trillion – see Bianconi (2013). There are, let us agree, trillions.

digestion. They make essential vitamins and other molecules. They break down toxins and metabolise drugs. They exert an invisible influence on our hormones, our immune systems and perhaps even our brains. And they crowd out other, potentially harmful organisms by filling the niches that they would occupy if they could. We would miss our many, many microbes badly if they were not there.

Scientists call the whole ensemble of microbes that make their living as fellow travellers with some larger organism the *microbiota*, or more often these days the *microbiome*.* My microbiome is as alive as I am. It develops, responds, adapts as life goes on, just as my own body cells do. So does yours. What that means for how our lives play out, and how we should think of the enlarged cellular community that constitutes a person, is just beginning to become clear – with the aid of techniques more powerful than any microscope.

---

* Some use 'microbiota' to mean all the microorganisms that live in some defined space, and 'microbiome' for the total mass of genetic material they carry. The latter word, coined by geneticist Joshua Lederberg, is usefully (I think) ambiguous. It could refer to a microbe/biome, in the ecological sense where a *biome* is an interacting collection of different species. The more recent assumption would be that it is made by combining microbe and '-ome' as in genome, proteome, and all the other '-omes' biologists now like to go to. This embodies the tension, or collaboration, between different biological disciplines that contemplate the microbiome. In practice, the terms microbiota and microbiome seem to get used interchangeably, and the latter now often displaces the former. Here, I'll use 'microbiome' in the sense of all the organisms, unless the context makes it clear we're just talking about DNA.

Show us your DNA

We have been looking at microscopic organisms for 300 years or so. But very recently we started looking at them in a new way. The first human genome – the complete set of genes in each of a person's body cells – was completely sequenced a bit over a decade ago. That achievement paved the way for routine, large-scale sequencing. There is still lots to work out about our genes and how they operate, but what we do certainly know is how to sequence large amounts of DNA. That has given us an amazing new window on to the microbial world.

Biologists used to work on microbes – mainly bacteria and viruses – one at a time. Most of what we know about DNA and how genes work comes from studies that began in bacteria, especially one that became a laboratory workhorse: *E. coli*. 'What's true for *E. coli* is true for an elephant' was a tongue-in-cheek slogan heard among the pioneers of molecular genetics in the 1960s, in recognition of how much they had invested in one small organism.

That research in turn was founded on good microbiological practice first laid down in the 19th century. Generations of 20th-century molecular geneticists worked with colonies of bacteria grown up from a single cell in a shallow dish layered with nutrient jelly. That is not how microbial life is lived outside the lab, and it works only for a minority of types of microbe. Bacteria, like us, normally live in a world that everywhere teems with other life. We knew that, in theory. But the new-style DNA analysis has been revelatory, especially in showing how varied and complex microbial life is.

Nowadays it does not matter if you have a pure sample of anything. Just take whatever mixed-up matter you can lay hands on, extract the genetic material and work on all of it at once. This is the new science of 'metagenomics'. It begins with taking a bunch of stuff that may contain living cells, or maybe viruses, cutting up all the DNA, and sequencing it. Seawater, soil, and shit are good things to sample. What usually comes out of this rough-and-ready analysis is a huge higgledy-piggledy list of genes. Then the researchers try to figure out what they all are, and where they come from.

This genetic window offers a startling new view of the complement of cells, human and microbial, that make up the person I see in the mirror. It is a quintessentially modern view. The kilo or so of bacteria in my colon, for example, are a considerable cell mass. They are also a great store of information.

How big a store? The answer comes as a jolt if you think that the self you can inspect in the mirror is you. Most of your genes do not really belong to you at all.

The Human Genome Project focused attention on our own chromosomes, the carefully packaged lengths of double-stranded DNA that reside in the nucleus of each human cell and help to define our individuality. They turned out to have 24,000 genes altogether. That was a lot fewer than the figure of 100,000 or so that was regularly quoted before we looked properly. Still, it appears to be enough to support a complex organism with trillions of cells divided into around 200 different cell types.

The way our microbes organise their genes is quite different from the one-size-fits-all approach that the cells in a large, multicellular creature adopt. First, the number of microbial cells is higher. Counting is hard, and calls for adding up population estimates for guts, mouths, noses and vaginas as well as sampling skin. Published figures for the number of bacterial cells we carry range from 30 to 400 trillion. If the total population of cells in a person decided what happened in our life by majority vote, the bacteria would probably win.*

But that is only half the story. How many microbial genes are there? Again it is hard to be precise. These microbial cells do not share a genome, and there are many different species involved. But the DNA tells us that the total number of genes in one typical human microbiome is around 2 million. That is a hundred times as many as we maintain in our own cells. Moreover, as you move through a human population the number of our genes stays the same. Every human microbiome is different, though, so the number of microbial species, and genes, keeps going up as more samples are analysed. The number of genes that have ever been registered in *any* human microbiome is now five times higher than in any single individual's complete microbial complement. Let me spell that out again. Human cells have 24,000 genes. All the microbes that live on and in human beings incorporate 10 million of them.**

Genes allow organisms to do things, and this is an enormously rich genetic resource. We are just beginning to find out what it can do for us. We already know that our personal load of bacteria help digest our food, process drugs, and activate our immune systems. They are involved in a raft of diseases, especially those affecting the bowel. Indirectly, they may affect whether we get fat, develop cancer, or even suffer high blood

---

* Many people cite 100 trillion, compared with 10 trillion body cells – but the bacterial figure is an old estimate based on analysis of a single gram of stool. The precise number does not matter much in what follows. See Smith (2014)

** When I drafted this section, the biggest catalogue listed 9,879,896 genes. See Li (2014). By the time this book went to print, the total had broken 10 million. See Karlsson, (2014).

pressure, heart disease or strokes. They seem to affect asthma. And there are hints that the precise make-up of the bacterial population in our guts can even affect brain development and behaviour.

This is just the start, though. Many of the microbial genes already found are of unknown function. We are still finding new species, new genes and new interactions. As well as bacteria in the microbiome, there are single-celled organisms of other kinds. There are numerous fungi that find humans congenial hosts. And there is a largely uncatalogued array of viruses, which add more depth to the genetic reservoir. We are covered in life, awash with it, saturated with it, in such variety that it is hard to take in.

As well as an impressive genetic resource, we are also finding that our microbiome is the wellspring of a vast, largely unmapped reservoir of human diversity. The range of species and strains of bacteria can differ wildly in different people. Even close relatives or those who live together maintain some microbial differences. And we all have different microbial populations in different parts of the body, from the armpits to the anus. They change over time, as we eat different foods, grow older, move from place to place, swab, scrub, or disinfect ourselves, or swallow antibiotics. For once, science really has revealed a new world. This one is in inner space. It is part of us.

What about me?

Sometimes science advances because of some radical conceptual breakthrough, from a Newton or an Einstein. More often, it moves ahead more slowly, as small observations alter the picture we are building of reality bit by bit. What is happening now is different again. It is one those times when a sudden leap in observational power transforms the view – and itself leads to a kind of scientific revolution.

That is exciting, fascinating, and, as the news filters through to the rest of us, a bit puzzling. I want to know what the emerging picture of the microbiome means for me. It seems amazing, when science recovers signals from the origins of the universe or probes the ultimate constituents of matter, that the profusion of life we carry, and have always carried, has largely eluded us for so long – hidden in plain sight, almost. If science is one of the ways in which we can know ourselves more completely, we have taken a long while to get round to this part of ourselves.

I am usually wary of stories about what this or that scientific revolution means for us. What do the latest findings in cosmology, particle physics or earth science mean for humanity – or even (and I am never sure what to make of this phrase) 'what it means to be human'? But the revolution in understanding the human microbiome is actually about us, and about me. It seems fair to ask what it all means, if not for what it means to be human then certainly for what it is like to be alive in the world.

There is no single answer. Exploring this world is changing our view of many parts of our lives. The flow of new scientific publications about our innumerable microscopic companions is a steady stream now growing towards a flood. More and more labs are joining in the work of finding out who's there and what they are up to. A crude measure gives one indication of how many. Google Scholar, which searches academic journal papers, finds just fifteen hits for 'microbiome' in 1995, and still a mere 30 five years later. There was a slight increase, to 76, by 2005, then an explosion of results and discussion:

2,190 papers in 2010, rising to 9,300 by 2013. At the moment, following microbiome research is not so much like keeping moving goalposts in view as tracking a rocket accelerating off the launch pad.

All this new research will change the way we tackle lots of problems, especially medical problems. It will change how we think of ourselves. We are still upright primates, who share a common ancestor with chimps and have chattered our way to a new kind of culture, technology and civilisation we are proud to call human. But now we look a little different. We have, some suggest, a whole new organ to look after – a microbiome. Alternatively, we are walking, talking bioreactors, wearing thousands of other species, and incubating thousands more in our guts. Then again, the whole assembly can be described as an ecosystem, or really a collection of ecosystems, all busily oper-ating at the cellular level. But there is yet another description on offer, which may best sum up what it is to have such a huge collection of microbial fellow travellers. It is a description that recognises that many of them are useful, if not essential. They are not commensals, as biologists refer to cohabiting organisms that simply do no harm (it means 'sharing a table' in Latin); they are full, mutually supporting partners, each relying on the other for mutual support. That kind of sharing has another name: symbiosis. Microbiome studies tell us that we have more symbionts than we ever dreamed. And the whole ensemble they compose, along with us, is a *superorganism*.*

---

* A term also proposed in this connection by Joshua Lederberg. Influential chap.

Taking a closer look

This book is my attempt to get to know the new, super-organismic me. That attempt leads down various paths. Superorganismic me is not completely new, so there will be a little history. Researchers are still working out how to study superorganisms, and getting to know my superorganism also means looking into how they uncover its workings. It also shows that focusing on the superorganism is changing science. Exploring the new world of the microbiome calls on many disciplines. Molecular geneticists, microbial ecologists, infectious disease specialists and immunologists are all having new, sometimes halting conversations with one another. All of them are having to talk to the new kids on the block, the bioinformaticians, who manage the databases and the software that help make sense of the microbiome as information. As these exchanges grow richer, the disciplines are changing in the course of the conversation. In particular, immunology is having its own conceptual revolution. Managing this new, carefully maintained consortium that makes up a superorganism turns out to be the reason our immune system evolved in the first place.

The rest of the book takes all this in stages, taking up a series of questions that follow from the big question: what does it mean to be a superorganism? Some of the individual answers are easy to get at, some are still emerging. But even where the details are still work in progress, we already know enough to glimpse the main outlines.

In Chapter 1, I start with an easy one. How did we first become aware of an invisible world of microbes? It is rather marvellous how an apparently simple instrument, the microscope, produced such an enlargement of our view of life, after hundreds of thousands of years during which humans remained oblivious to the smaller dimensions of biology. The first reaction, not surprisingly, was of wonderment. Then heroic scientists invented germ theory, which saved millions of lives but did microbes' reputation no good *at all*.

Chapter 2 considers who we are actually sharing our lives with by asking what microbes can do, apart from cause disease. It turns out that even the simple ones, bacteria, can do almost everything multicellular life can do (OK, they don't write books). Oh yes, and they run the planet, just as they always have done.

Then Chapter 3 asks how we know what we know about the microbiome. Why, for one thing, did it take us so long to realise that it is a vital part of human life? Partly, it is a matter of technique. The dazzling advances in DNA sequencing and genetic analysis are the main things that now afford a new view of the life within us. Now we know it is there, researchers have to devise new methods to study what it actually does, using experimental systems including germ-free mice carefully resupplied with intestinal bacteria in known combinations, artificial ecosystems living quietly in bench-top flasks, and bits of cultured intestine for the bugs to grow on. Many variations on these in a host of labs are helping us build up the new picture of the life on us, and how it lives.

What does this new observational arsenal tell us about who, and how, 'we' are? Chapters 4 and 5 ask, simply, 'who's there?' – with the answers coming mainly out of DNA analysis. They vary between people, and depending on where you look. We can think of a complete microbiome, or many microbiomes: on the lips and teeth, on each finger, behind the knees, in the belly button, armpit and groin. I take a look at some of the well-studied sites – the mouth, the skin, the vagina, the lungs. Then in Chapter 5 I take the measure of the lusher pastures of the organ that contains by far the largest microbial community, the gut. This, if anywhere, is the heart of the superorganism.

Now we know something of who we are sharing our lives with, Chapter 6 asks: how did they all get there? There are two different answers to explore. There is the story of how these microbial communities get established in each of us. How is the human microbiome born and how does it develop? And there is a much longer story that we can piece together relating how humans, and all their ancestors, evolved with their microbial accomplices in on the act. What microbiomes do other primates have now? What might our proto-human ancestors have nourished in their guts? That leads to a question that is worrying some microbiome researchers: what changes has modern life, with surgical delivery of babies, antibiotics and fast food, brought to our microbial make-up? Is it possible that we are discovering the really important features of our microbiome at the same time as we are messing it up?

Chapter 7 looks more closely at how all these microbes interact with our own cells, tissues and organs. The immune system is at the heart of this interaction, and it turns out that for us to have useful ideas about how it relates to our microbiome we have to think about it in a new way. The idea that immune cells wage constant war on foreign agents of disease has dominated scientific and popular talk about immunity for decades. How can we reconcile that with the fact that trillions of microbes live inside us? A subtler, constantly shifting, but much more carefully negotiated relationship with microbes now looks like a more realistic picture. This shift in our view of this complex aspect of human being is the most important scientific development arising from microbiome research so far.

Negotiations can be subtle, but still break down. Chapter 8 asks what happens when our microbial colonists tear up their tenancy agreements and set about trashing the joint. The medical effects of alterations in microbial populations are many and varied, and not all well understood, but there are strong pointers to their importance. Changes in our microbiome can be charted in bowel disease, autoimmune conditions, cancers, obesity and more.

Many of the medical conditions linked with the microbiome originate in the gut. But what else can it affect? In particular, can these tiniest components of our superorganism influence the seat of reason, the brain? Chapter 9 moves into this more speculative territory, where the microbiome and the self truly interact, and asks how microbial influences on brain and behaviour might happen, and what we know about possible links with some mental illnesses such as depression.

Bacteria may rule the planet, but they have help. Not from multicellular organisms, but from still simpler entities – viruses. It is estimated there are ten viruses for every bacterium, wherever they are found. Our bodies are no exception. So what do we know about this next layer in our newly uncovered microbiome? Not so much, but Chapter 10 looks at what we do know. Investigation of our viral community is just beginning but may emerge to be just as important as the rest of the microbiome.

It looks as if our microbiome has changed in modern times. Now we know so much more about it, we will probably set about changing it again. Chapter 11 asks what we might decide to do with it. Should we try to restore it to some happier primeval state, or will we be more inventive and refashion our personal ecosystems to suit ourselves? The two methods now on offer are drastic: a faecal transplant to repopulate your bowel; or – gentler but not terribly effective – eating probiotics like yoghurt or sauerkraut. Both can be improved. We already genetically engineer lots of bacteria to suit ourselves. What are the prospects for a designer microbiome later this century?

That leaves just one question for the final chapter – the one I started with: what does it mean to be a superorganism? I cannot offer one big answer to my big question, so I give you lots of small ones. There is some advice, drawn from research, on how to look after your microbiome. I look at where the research is likely to go next, with some as yet unanswered questions and predictions for how answers might be teased out. And I go back to the mirror, to contemplate my new, composite self.

Whichever way we look at it, our bodily life is intimately involved with myriad other small lives. Reading genomes has already revealed a new dimension of our connection with the rest of life, in the genes we share with every other organism. Getting to know our own microbes shows yet another. Will we end up valuing them more, even admiring them? In a world of antibiotics, antisepsis, disinfection, and pasteurisation, will we try to look after them better? Or can a superorganism look after itself?

But before I try to take stock of present-day microbiome science, let's go back to our very first direct encounter with microbes; an event so recent in history we can date it rather precisely.

1 | Strange new world

In 1676 Antonie van Leeuwenhoek, a prosperous Dutch draper, was ten years into one of the most breathtaking observational binges in the history of science. A decade earlier, he pored over the images in Robert Hooke's celebrated *Micrographia* (1665), marvelling at its mind-expanding drawings of things never seen before – the multifaceted eye of a fly, the barbs on a bee sting, a louse clasping a human hair.

Inspired, Leeuwenhoek perfected a way of making simple handheld microscopes with minute spherical lenses. He soon saw what Hooke saw, but his superior instruments allowed him to go further. Yes, large organisms had extraordinarily intricate small parts. But there was more going on beneath the level of normal human vision. In a drop of pond water he reported seeing minute single-celled creatures, 'animalcules', as he called them. He also saw even smaller creatures, the first of them in water in which he had ground up pepper in an effort to investigate its spicy taste. His claims, included in a 1676 letter to the Royal Society in London – then the best way of informing others of new scientific findings – seemed bizarre, but Hooke returned to his microscopes and confirmed it was true. When the Dutchman's findings were published in the

Society's *Proceedings*, Leeuwenhoek's animalcules were a sensation – the first entrancing glimpse of microbiology.*

Seven years on, and one of his now regular letters to the Royal Society described what he saw when he scraped some of the white matter from between his teeth, 'as thick as wetted flour', mixed it with rainwater or spittle, and examined it at high magnification. 'To my great surprise ... the aforesaid matter contained very many small living animals, which moved themselves very extravagantly'.

Motion was, for Leeuwenhoek, the sign of life. He assumed that the many small bodies he saw that did not seem to move were dead matter. Now we know they are just less mobile microbes. But his description of his miniature zoo was still a revelation. There were new kinds of life, invisible and unimagined in all earlier human history. And they were not just out there in the world but living on us.

The insatiably curious Dutchman was full of wonder at the life he had excavated from his own mouth. He described with delight the animalcules' 'pleasing motions', the ways they moved, their variety of shapes, their sheer profusion. The smallest kind had motions that reminded him of 'a swarm of flies or gnats flying and turning among one another in a small space'. They were so numerous that, he wrote, 'I believe there might be many thousands in a quantity of water no bigger than a [grain of] sand'.

He found the same inhabitants in the mouths of two ladies – probably his wife and daughter – who, like Leeuwenhoek,

---

* Strictly speaking, we must credit Hooke with recording the first glimpse of a microbe: a fungus, which he described as 'small and variously figured mushrooms'. See Gest (2004). Van Leeuwenhoek was the first person to describe bacteria.

cleaned their teeth regularly. Elsewhere he told how when he scraped the teeth of 'an old gentleman, who was very careless about keeping them clean', he found, 'an incredible number of living animalcules, swimming about more rapidly than any I before had seen, and in such numbers, that the water which contained them (though but a small portion of the matter taken from the teeth was mixed in it) seemed to be alive.' Urging his main conclusion on the Royal Society, he reached again for the right comparison. 'The number of animals in the scurf of a man's teeth are so many that I believe they exceed the number of men in a kingdom'.

The revelation that we teem with other life was a sensational exhibit in a catalogue of wonders that the microscope made visible. Along with the telescopic discoveries of Galileo, natural philosophers' new access to the micro-world was the first thing to establish that science could only gain by using new instruments to go beyond the unaided human senses. At first, not everyone could accept that phenomena hidden from normal vision were real in the same way as those that pass the simpler naked-eye test of 'seeing is believing'. But the majority believed that science was gaining unprecedented new access to important knowledge about things in the world. In this way, our own microbes had a starring role in the genesis of a recognisably modern way of doing science. That makes it more surprising, somehow, that so much about them remained unknown until so recently.

A molecular menagerie

A little over 300 years after Leeuwenhoek, another curious human scraped his own teeth in search of wildlife. To be accurate, David Relman got his dentist to do it. Instead of discarding the gunk from Relman's gum crevices when he cleaned his teeth, the dentist put it in sterile collection tubes that the Stanford University researcher had taken with him to the surgery.

Relman had been getting to know new DNA-based techniques to pin down pathogenic bacteria that resisted identification because they refused to grow in culture in the lab. He got to wondering if many species were going overlooked in the complex population mixtures of our normal microbiota for the same reason. Back in the lab he followed normal microbiological routine and set up cultures from the samples. But the key results came from an addition to the routine.[1] He used some of the sample for the latest DNA analysis, seeking small pieces of gene sequence characteristic of bacteria and comparing them with known sequences in scientific databanks.

It would have been easy to assume there was nothing much more to find in the pockets between teeth and gums, the sub-gingival crevices. Over the years, careful bacterial cultivators had logged almost 500 different species of bacteria recovered from this well-populated region of the mouth.

However, working over this one-shot sample from two teeth in one mouth, Relman's team found 31 new strains of bacteria, identified by their DNA sequences. Another six turned up on the culture plates for a final tally of 37 new kinds – out of 77 in total. Probing bacterial DNA uncovered a whole new dimension of life on us.

In the sober language of the *Proceedings of the National Academy of Sciences* in 1999, Relman and his colleagues reckoned that 'Our data suggest that a significant proportion of the resident human bacterial flora remain poorly characterized, even within this well studied and familiar microbial environment'. Or, as he later told the *San Francisco Chronicle*, 'We found much, much more with the molecular methods than we found with cultivation. That meant we'd been missing this huge fraction of the microbial world for more than 100 years. That's a humbling thing. We were playing with half a deck.'

That realisation fuelled a big effort to apply the new technologies of DNA sequencing to microbial samples from as many human body sites as people could poke, scrape, rinse or mop up. Since the millennium, the results of this effort have transformed our picture of the human microbiome, and of how we and a myriad of other species coexist.

But before we look any more closely at the results of these deeper probes into the unseen world, and the questions of meaning with which scientists are now grappling, let us go back. Because in the three centuries between these two enterprising observers of teeth, we learnt a few other things about microbes.

Good guys, bad guys

Leeuwenhoek's animalcules were fascinating to enlightened society, but seemed mainly a harmless novelty. Some simple experimentation showed that the newly fashionable 17th-century beverage coffee, or a little vinegar, destroyed the life in his field of view. Besides, the idea that creatures so small could have any important effects on their hosts seemed fanciful.

Now we know better. The most significant changes in knowledge came in the 19th century. The germ theory of disease emerged from a combination of a closer investigation of infection with a newly systematised science of microbiology.

Contagion, or close contact, had long been associated with the spread of some diseases – but contagion with what? Now it became clear that the crucial contact was with microbes, and it was thus convincing to claim that microscopic life had momentous effects on much larger organisms, with microbes as the main actors in a compelling new explanation of some deadly illnesses.

That also had a big effect on how people thought about microorganisms – two kinds of effects, in fact: scientific and cultural. Both are still very much with us.

Scientifically, part of the legacy of germ theory is a template for how to reason one's way through causes and effects in the moist, mixed-up world of biology – a template for microbial logic, if you like. It still conditions a lot of our thinking about links between the human microbiome and disease, although, as we will see, it is much harder to apply to the kind of results modern research delivers.

Back in the 19th century, the mysterious organisms that showed up under the microscope were mostly recovered from outside us. Finding them inside mainly happened when people investigated disease. But were the minute creatures recovered from patients or sick animals (and perhaps kept alive as cultures in the lab) causing the symptoms? It is hard to recover this mindset now, but there were reasons to doubt it, and plenty of sceptics. Persuading them that the theory was sound demanded a mass of evidence, and then some clear rules of inference. Then you could build a watertight case.

The germ theory codified those rules. At its simplest, the theory assumed the form of the 'OGOD' hypothesis – One Germ, One Disease (this later spawned a close relative, One Gene, One Disease, but that's another story).[2]

If OGOD is true, and there are lots of germs about, how can you tell you have found the guilty organism? The rules derive from a general approach to scientific experiments that we now take for granted. It was first described formally in the 19th century by the philosopher John Stuart Mill, who called it the method of agreement and difference. It is the recipe for the perfect experiment that we learn in school. Define *all* the conditions in some controlled set-up. Vary them one at a time, and see what happens. If variable X causes a change in result Y, then the two are linked in some way. The easy example is working out what caused some of the party to get food poisoning after dining out, by detailing who did, and who did not, eat various things.

For germs, the details came from the German bacteriologist Robert Koch (1843–1910). In the early 1880s, he and his great rival Louis Pasteur had established connections between a few diseases, mostly in animals, and specific infectious agents. Koch wanted to generalise, and to quiet doubters.[3] Along with ferociously energetic lab work, and advances in method (he pioneered both the use of microbiological plates instead of flasks of broth for growing colonies, and staining bacteria with dye to aid microscopic identification), he formulated the rules known as Koch's postulates. There were just four. They translated easily into instructions for demonstrating that a germ really caused a disease. Do these four things and you could be sure you had the answer, and convince everyone else:

1. Find the microorganism in *all* the subjects (animals or people) who have the disease, but not in healthy specimens.
2. Isolate the microbe from a diseased organism and grow it in the lab, in culture.

3. Dose a healthy host from the culture, and give it the same disease.
4. Isolate the microbe again from the newly diseased host, and show that it is the same as the one you started with.

None of this was exactly easy, even in experimental animals, let alone human patients (step 2 was especially frustrating). But as methods improved, the four postulates proved their worth, advanced science and earned the gratitude of millions. The great scourge tuberculosis was the test case for these rules. Cholera, typhus, tetanus and plague followed and were all correctly identified as infectious diseases in the next dozen years.

The logic remains sound, provided it is the bug, and only the bug, that is involved in the disease. Many later cases – and quite a few of the classic ones, like tuberculosis – are a good deal more complicated than that. But it remains the often-cited gold standard for working out the links in chains of cause and effect that lead from other organisms to effects on people. Is it helpful in unravelling cause and effect in the complex ecologies of our microbiome? We will have to come back to that question later.

The power of Koch's logic, though, reinforced the cultural impact of the germ theory. Along with spectacular medical successes, the newly white-coated microbiologists of the 19th century were also illuminating the beneficial roles of microbes. Pasteur, in particular, was interested in fermentation as well as infection. But the fanfare that accompanied demonstrations that microbes could cause disease tended to drown out the news about the good guys that were making cheese or wine. The idea took hold that germs, with a few honourable exceptions, were evil.

Don't touch – dirty!

Here's how to open a can of peaches. Remove the label, then scrub the can to remove all traces. Open the lid, and pour the peaches into a bowl. Do NOT let the can touch the bowl.

So staff were instructed in the kitchen of Howard Hughes, pioneering 1930s aviator, film-maker, billionaire recluse and long-time sufferer of obsessive-compulsive disorder.[4] The same staff had to wash their hands until they were sore, and wrap them in paper towels when they served Hughes' meals. There were detailed instructions on how to open the packaging for the towels.

Hughes' deteriorating mental condition made him an extreme case of a common dread. He was fabulously wealthy, but had brain injuries from several air crashes and had contracted syphilis as a young man. Before any of these things happened he also acquired a lasting fear of germs. His later sad fixation on their dangers is emblematic of a culture preoccupied with an insidious microbial threat to health.

It is easy to see why. The germ theory of disease came when city dwellers were suffering from infections that spread through populations crowded together in unsanitary dwellings. The theory was a colossal success. It won over scientists when Koch, Pasteur and others showed that much-feared illnesses really were caused by tiny organisms. It won over the public by being easy to understand, and because – via vaccination and building proper sewers – it paved the way for their prevention and treatment. Some feared diseases were even eradicated. It remains a cornerstone of the new discipline of public health.

It was, in short, a scientific and medical triumph and the scientists who established it were heroes. Paul De Kruif's classic

The Microbe Hunters, a 1920s popular book by a writer who spent time at the Rockefeller Institute for Medical Research in New York, depicts a series of bacteriological pioneers in that light. It caught the tone of innumerable newspaper profiles and a clutch of later biopics.

They were heroes because they waged war against disease, and triumphed. And germs were the enemy. The celebration of science cemented a powerful association between germs, dirt, and disease. And the practical uses of antiseptics, disinfectants and antibacterials were impressive.

So avoiding germs is not just a matter of eschewing obvious sources of unpleasantness – don't smear shit on the walls, don't eat spoiled meat. It calls for dealing with threats that we cannot see.

This linked set of ideas has great power. Germs cause disease, come from dirt, and can spread through contamination you won't even notice. It contains a core of truth. Some germs really do make people sick. You really can't see them. And strict precautions will help prevent harm. You really should use the hand sanitiser when you visit the hospital.[5]

It has also been advertised endlessly, in public health campaigns and by companies making anti-microbial products. They are always on the lookout for new things to disinfect – toilets, bathrooms, kitchens, babies' bottles, food packaging, telephones, keyboards and, not least, people.

The course of this never-ending campaign is laid out in historian Nancy Tomes' 1998 book *The Gospel of Germs*. In the US, public health advocates set the early pace, urging a new standard of cleanliness and hygiene on everyone, but especially housewives and mothers. Later, medical wisdom altered to put less emphasis on sheer cleanliness, more on avoiding contact with infection and preventing those infected from spreading a disease organism. So industry took up the slack.

It was everyone's duty to wage war on germs, but women were still in the front line. A US Lysol advertisement from 1941 hammers the point home. It seamlessly blends domestic with patriotic duty with a picture of a smiling housewife throwing a military salute. 'You're in the army, too,' runs the ad. 'Enlist now for the war on germs ... A woman with a mop, a pail of water and a bottle of "Lysol" can rout an army of bacteria that cause infection ... [Lysol] is the housewife's home defense.'*

Disinfectant liquids, sprays and wipes are still big business, as a glance down the supermarket aisle shows. Lysol disinfect-ant spray now comes with ten different scents, but the main selling point remains that it allows you to 'protect your family from germs they could come into contact with every day'. By killing them.

Many of us in the West have probably become a good deal more casual about using such products in the home, except in the lavatory. I certainly have. But we still recoil from public spaces that are obviously unhygienic. And it only takes the fear of a new or re-emerging disease to bring back old ideas about contagion and contamination.

---

* Not all Lysol ads are so easy to read. Some pre-war magazine ads read like a serious assault on the microbiome, suggesting that Lysol 'truly cleanses the vaginal canal even in the presence of mucous matter. Thus Lysol acts in a way that makeshifts like soap, salt or soda never can. Appealing daintiness is assured, because the very source of objectionable odour is eliminated.' And, women, you need to do that to 'keep you desirable'. It should be said, though, that this ad is less about treating undesirable odour, more a none-too-heavily disguised advertisement for Lysol as a spermicidal douche.

Tomes ends her history by recounting the reaction to a case early in the AIDS epidemic, in 1984, in which a thirteen-year-old boy became HIV positive after regular blood transfusions for haemophilia.

When news of the diagnosis got out in Kokomo, Indiana, townsfolk refused to shake hands with him or use a toilet he had visited, spread rumours that he spat on vegetables in the greengrocers, and made him and his family sit in a pew out of cough range in church. The family eventually left town when the boy's decision to return to school brought a bullet through the living room window. Fear of contagion can still turn law-abiding citizens into vigilantes.

Microbes in the lab

While the public wanted to have as little to do with germs as possible, microbial life came under more intense scientific scrutiny. Over the next century, scientists who were happy to get up close and personal with microbes learnt an immense amount about the varieties of minuscule life, and found new ways of probing their inner workings.

That involved adding to the simplest aid to observation, the microscope, with new ways of preparing and staining cells to enhance details, learning how to culture microbes for longer study, and investigating their chemistry and, eventually, their genes.

As microbiology developed in the time after Leeuwenhoek, slowly at first then much more productively from the mid-19th century on, many more subtle techniques came into use. The catalogue of microbes expanded enormously, and more and more species were found in more and more places, from hot springs to ocean depths.

The microbes living with us were part of this steady effort, and it was occasionally suggested that as they are our constant companions, and are so numerous, the vast majority probably do us no harm, and may even do good. (Pasteur had in fact suggested this back in 1885, but the message went unheard then.[6])

This view surfaced again many years later, but before the era of DNA analysis, in a prescient book by the 1960s authority on humans and their microorganisms, Theodor Rosebury. After writing a textbook on the subject, he produced a quirky popular book in 1969 called *Life on Man.** It is intriguing now partly for the growth in knowledge we have seen in the intervening decades. This was a summary of the state of the art by the man who owned the subject. But apart from describing some species and their prevalence, he had very little to say about the details of the microbiome, especially in the gut. He regarded most of the mass of bacteria we carry as harmless passengers, with a few side-benefits like discouraging colonisation by organisms that can cause disease. He mentions some other suggestions, such as microbes' possible influence on development of the intestines, but only as matters that deserve further exploration. His summary of important things to know about actual life on humans runs to a scant 24 pages, with the rest of the book taken up with an entertainingly erudite discussion of the anthropology of disgust. It is less than a lifetime ago, but scientifically a totally different era.

---

* I see my copy dates from 1976, so I'm giving myself credit for finding the topic interesting back then.

Nevertheless, the wider science of microbiology had learnt a great deal about the crowds of smaller species that are part of the living world. As we try to build a picture of what it is like to share a life with a large cohort of microorganisms, let us take a closer look at how they live, focusing on the bacteria.

2 | Microbes aren't us – or are they?

Consider a single bacterium. Let us make it *Escherichia coli*. It is briefly airborne and drops on to the jellied surface of a warm, freshly prepared culture dish, rich in nutrients. It can do a variety of things, but its absolute top priority is turning into two bacteria. In these ideal conditions, with room to expand and no competition, that takes as little as twenty minutes.

In that time, it copies all the DNA in its single chromosome, and manufactures enough proteins, cell wall scaffolding, and other cellular constituents to double up everything it needs. Then it divides into two cells, each identical with the original. Twenty minutes later, each of the new cells divides again.

Unconstrained, in a culture dish the size of a planet, this will go on happening. The microbe is so eager to reproduce that the newly duplicated chromosome is already starting a new round of replication before cell division happens. Otherwise the enzymes copying the DNA could not do the job fast enough.

The daunting power of exponential growth leads, mathematically at any rate, to impressive results. In seven hours, there will be a million *E. coli*, still dividing. In *The Andromeda Strain*, Michael Crichton has one of the characters say, 'It can be shown that in a single day one cell of *E. coli* could produce a super-colony equal in size and weight to the entire planet Earth.' This is wrong, though: it would take nearly two days.

That never happens because *E. coli* exhausts the food supply and begins to choke on its own waste. It responds by abandoning division and going into a stationary phase, with many systems shut down, or merely ticking over, until another chance arises for glorious growth.

Speed of reproduction down among the microbes is one of the things that makes them so adaptable. There are many thousands of different bacteria. In any one place, whether it is a deep ocean vent or deep in our guts, the conditions will suit some of them. Others will still be around in much smaller numbers. For now, they have been outdone by their better-adapted competitors. If conditions change, perhaps they will get their chance for an exponential splurge.

The two extremes of fast reproduction and near-complete dormancy help make bacteria versatile, and ubiquitous. They can live practically anywhere, on more or less anything. They have been here practically for ever, and tried every possible metabolic trick. Even if none of them were interested in living on us, it would be worth knowing more about them just to understand the main part of the story of life on Earth. But their story is not separate from ours. It is tempting to think of them as very different. But we have much more in common with our dominant microbes than we once thought.

Dominant species

The human microbiome has a bit of everything, biologically speaking. A large beast like *Homo sapiens* can provide a home for lots of other creatures, welcome and unwelcome. Our attitude towards them is influenced by a selection of parasites and verminous passengers that are too large to be part of the micro-biome – from lice to tapeworms. Focus on the micro-level, where the numbers are so much greater, and there are many of the very simplest life forms: viruses (of which more later). There are also representatives of relatively complex life: single-celled eukaryotes (or *protists*) with their apparently more organised cells tricked out with nuclei and other sub-cellular structures visible under the microscope. But the biggest contributors to the microbiome, measured by both total mass and numbers of genes, are bacteria and the partly similar archaea. They are also the dominant influence on the rest of the organism.

We tend to think of them as pretty simple, something between a curiosity and a nuisance. True, they are prokaryotes* – that is, they do not parcel their DNA up into a cell nucleus but have a single, circular chromosome and usually some other assorted bits of DNA in the main body of the cell. Indeed, the main body of the cell is all there is. Under the microscope, at low magnification at any rate, they *look* much less sophisticated than the single-celled eukaryotes, with their apparently much more elaborately structured insides. And it is eukaryotes alone that led to the highly evolved multicellular organisms, like us, that we are more likely to notice.

The apparent simplicity of bacteria seems to suit their small size. Bacteria are really *micro*-microorganisms. A typical prokaryote is thousands of times smaller in volume than an average eukaryotic cell. It has a genome around 10,000 times

---

* Prokaryote and eukaryote are possibly the two most annoying terms in life science. Yes, the words are very similar. It is damnably hard to remember which is which. But the distinction does mark probably the most important division in kinds of life. I apologise on behalf of biologists everywhere.

smaller. It seems, superficially, less *organised* with what internal structure there is remaining unseen until the introduction of high-powered electron microscopes. For a long time, the inside of these tiny, single-celled organisms was thought to be basically a reaction vessel. A bacterium was a collection of bioactive chemicals in a bag.

They can seem less interesting if we take the long view, too. The evolutionary story is often told as one of steady progress, beginning with primitive microbes and culminating in smart primates. Bacteria did some heavy hauling around the origins of life but their true destiny was realised only when they somehow made the leap to nucleated cells, then multicellularity, then the *really* interesting life forms, with brains. But they are much more than a hangover or a sideshow. However life began, for 2 billion years bacteria and the other single-celled division of life, the archaea, were the only game in town. With a generation time measured in minutes, they had time to try out unimaginably many evolutionary options.

More important, they were not superseded by more complex life. That is not how evolution works. They continued living and evolving in their own niches, which finally included other, larger organisms. They remain, from many points of view, the dominant life forms. Their total biomass is similar to that of all the plants and animals on Earth. Until one advanced primate began to disinter fossil fuels and set fire to them to fuel a civilisation, they were also the main way that life influenced the environment on a planetary scale.[1]

This view of the significance of bacteria for the biosphere as a whole has emerged in the last few decades, especially as they have been found in more and more extreme environments. There are two other key dimensions in the expansion of our knowledge of bacteria. Both are important in thinking about them as part of our extended cellular community. We now have a far more sophisticated idea of how they live, and what they can do. And we have a deeper appreciation of what we and they share; of why – as biologist and essayist Lewis Thomas once pointed out – it seems strangely prescient, linguistically, that *humus*, made by trillions of bacteria in the soil, and *human*, have the same root.

Small is *not* simple

What matters about bacteria? First, they *are* microbes; that is, really, really small. Your typical bacterium is one to a few millionths of a metre on its longest axis. That makes them easy to miss. For the vast majority of our own relatively brief time on the planet we had no idea they even existed. As we've said, they make up for individual minuteness by proliferating hugely. Plausible published estimates suggest there are something like 10^{30} – a million, trillion, trillion – bacteria on the whole Earth. Our ignorance of large swathes of that population is pretty deep. We may never know how many there are in total, or even how many different kinds.

On the other hand, now we know they are there, their small size and rapid growth makes some of them easy to study, one species at a time. So our comparatively thin information about the global bacteriosphere goes along with astoundingly detailed knowledge of a few much-studied organisms, and one in particular, the laboratory pet we keep coming across, *E. coli*.

Get to know *E. coli* a little and you do not get acquainted with all bacteria, but you do get a healthy respect for what bacteria can do.[2] Yes, they can grow and reproduce. That is what makes them alive. They have the full complement of tiny nanomachines for making copies of their own DNA, reading the information it stores, and translating it into protein molecules. They can digest food molecules, extracting energy from their breakdown and using the broken bits to build up new molecules that they need.

Much of what we know about these processes, from the details of the genetic code (the same in every organism on Earth) to the network of chemical transformations that form the basics of metabolism, was worked out from experiments on countless colonies of *E. coli* in laboratory dishes. But that is only part of this tiny organism's contribution to science. Many further experiments, often in conditions more like life in the wild than a single species growing in a dish, have shown how much else bacteria can do.

They have keen senses, at the molecular level. They don't see or hear, but *E. coli* and other microbes can detect changing concentrations of important molecules around them. They can move under power, propelled by a tiny, busily spinning flagellum like an overactive tail. They alter course to get closer to molecules they like (food) and distance themselves from ones they don't. They adapt to their environment, recognising changes such as temperature and the availability of particular foodstuffs. The response leads to switching genes on or off, with the switching organised in complex circuits that exploit the way one carefully tailored molecule binds to another. And they respond to other cells, through a phenomenon called 'quorum sensing', in which certain functions are activated only when the population density reaches a threshold. They wage chemical warfare on other bacteria, or assume cosy metabolic relationships with them in which one species consumes the partly processed molecular food already used by another. Often, they congregate in great cellular assemblies. While not achieving multicellular life, they get together to do things that look awfully like it, secreting sticky molecules that hold together a slimy 'biofilm'. Such films coat a surface that offers a habitable niche, like your teeth, and support an enduring bacterial system with a complex biochemical division of labour.

In addition, as Joshua Lederberg showed back in the 1940s in the work that won him a Nobel Prize, they have sex. *E. coli* and other bacteria are perfectly capable of reproducing without anything resembling sex – they can generate clones of genetically identical cells, notwithstanding the odd mutation. But they are also keeping their options open. Every now and again, two cells come into conjunction and DNA passes from one to the other. These gene swaps make the microbial world look very different from the clearly delineated species we recognise among multicellular, eukaryotic organisms. There is a constant exchange of genetic titbits, through transfer of portions of the bacterial chromosome, through movement of the small loops of DNA called plasmids that are also found in most bacteria, and via bacterial viruses. If none of those happen, a bacterium can even take up unaccompanied DNA from its surroundings, some of which may be incorporated in its chromosome, a process known as transformation. Bacterial DNA also typically changes faster by mutation than that found in the chromosomes of other organisms. This is not just because they can reproduce so fast. Microbes under stress – when food is in short supply, for example – copy DNA less accurately and repair it less well. Is that simply a by-product of stress, or is it a clever evolutionary trick to throw up lots of possible metabolic answers to a new problem? Biologists are still debating that, but either way this hyper-mutation allows rapid change.

Over time, all this gene swapping and DNA alteration leads to wide divergence between strains of bacteria that all look the same, and all carry the same species label. But close study of the bacteria we have come to know well shows that a 'species' down among the microbes is a perplexingly fuzzy entity. One kind of bacteria can draw on a large genetic resource, much of which is normally kept in reserve. As usual, its starring role in the history of biology means *E. coli* is the pre-eminent researched example. Different strains have a core of genes in common, but as more strains are analysed in detail, these are increasingly outnumbered by genes that are found in some *E. coli* strains but not others. As Carl Zimmer writes, 'the list of genes shared by every *E. coli* is getting shorter, while the list of genes found in at least one strain is getting longer.' A typical *E. coli* strain, if there is such a thing, has 4,000–5,000 genes. But the total of all the genes ever found in *E. coli* – dubbed its pangenome, though we could just call it a gene pool – now stands at 16,000. That is not far short of the number of genes in the human genome, an impressive total for such a 'simple' organism.

Humanoid bacteria, and bacterioid humans

So bacteria are more complex than they may appear, and as versatile and resourceful as befits a manifestation of life that has survived for 3 or 4 billion years. And as our predecessors and now contemporaries, their lives and evolutionary histories are bound up with our own in innumerable ways that we have only recently appreciated.

One of them is the fact, uncomfortable to some, that motivates this book. They live inside us. *E. coli*, the great exemplar, is no exception. The first samples were isolated in 1885, by Theodor Escherich, from newborn babies' early bowel movements. It was easier to isolate than the majority of other gut bacteria because it can live with oxygen as well as without it. A range of innocuous *E. coli* strains normally live in our colon. They are beautifully adapted to the guts of warm-blooded creatures. On the other hand, there are an equally diverse set of strains that cause unpleasant symptoms, such as food poisoning, or worse.

Then there's a much stranger kind of cohabitation that scientists found hard to believe for many years. We know that the entire biosphere rests on a base of single-celled prokaryotes. But it turns out that this is true not just in the sense that bacteria were here first and are still key players in many chemical and biological cycles. More complex life forms like us, with all the extra bits and pieces of the nucleated, eukaryotic cell, actually carry the descendants of ancient bacteria inside them.

Our highly organised, large-volume eukaryotic cells have a much richer energy supply than prokaryotes, even when you correct for their size. A startling reframing of cellular evolution helps account for how this came about.* They get their energy by using the output from the intracellular powerhouses called mitochondria. These look a bit like bacteria. That is because that's exactly what they are, or were. They long since lost the ability to live independently, but they still have a small genome of their own, which codes for – among other things – a DNA copying and reading apparatus that resembles the

---

* The appearance of complex cells may have been an extremely unlikely event – hence the 2 billion-year stretch when bacteria were the only living things. The way the chemistry and physics that shaped the energetics of cell evolution made the emergence of eukaryotic complexity so unlikely is explained in Nick Lane's brilliantly argued book *The Vital Question*, which ought to be published by the time you read this.

bacterial machinery rather than the quite different macromolecular marvels that do the job in the cell nucleus.

The explanation, offered by the late Lynn Margulis in the 1960s, is that a bacterial dalliance billions of years ago produced a symbiosis that tilted towards one bacterium living inside another. This internal colonist then adapted to getting all its sustenance from the surrounding cell, in return for releasing the energy it derived from chemical breakdown of sugars by oxygen. The result was the specialised organelle (that is, like a very small organ) in the fully fledged eukaryotic cell that acts as a miniature powerhouse – the mitochondrion. What was once a bacterium became first an intracellular parasite, then a rather simpler bag of infolded membranes that is dedicated to energy production.

Margulis saw this irreversible co-operation, or endosymbiosis, as a key evolutionary mechanism, and argued that several other parts of eukaryotic cells have the same origin. That remains controversial, but it is pretty much undisputed that both mitochondria and the chloroplasts that do a similar job in plant cells arose this way. It is a strange thought that all our cells are host to these ancient tenants, sometimes numbering in the thousands, and still dividing and reproducing independently. All of us are kept alive by an enormous collection of degenerate bacteria.

Finally, there are bacterial remnants doing another essential service in every other kind of cell. This is simply a consequence of the history of life, and of evolution by descent. All of Darwin's 'endless forms most beautiful' once had a common ancestor, and it was probably quite like some still-existing bacteria. That ancient progenitor had already acquired many essential genes – and it is often a feature of the really important jobs that proteins do in cells that the proteins, and hence the genes that preserve the information for making them, change very little in the course of evolution. Once they are up and running, any change via DNA mutation tends to make them work less well, and so the change disappears again as natural selection does its relentless winnowing of new ideas.

We knew this in principle, but the recent prodigious feats of decoding that make available for inspection the complete genomes of organisms large and small have reinforced how important it is, and how closely interrelated is all life on Earth. Compare the detailed sequences and it turns out 37 per cent of human genes have very similar counterparts in bacteria. That suggests they had already evolved in a common ancestor more than 2 billion years ago. That compares with 28 per cent of our genes that are just shared with other eukaryotes, 16 per cent shared only with animals, down to a mere 6 per cent that we can infer evolved in a common ancestor of us and other primates.[3] So whatever contribution our cohabiting bacteria make to our lives now, in evolutionary terms more than a third of our genes are among the things bacteria gave us.

Naming names

We have learnt a lot about bacteria, to the point where assembling a clutch of impressive generalisations like those above is fairly easy. Writing about them in more detail can be hard. Not because of their inherent complexity, especially when many kinds are jumbled together – although there is that. No, it is mainly the names. They are a mess. As I will have to use more of them – we need to move on from *E. coli* – a word of explanation as to why they are a poor advertisement for science.

We know now that a bacterial species is not quite the same

thing as the separate strand of life that the term denotes in other organisms. (Actually, defining a species in the other kinds of organism is not that straightforward either, but let that pass.) The naming system for bacteria, though, follows the same conventions as biologists use elsewhere. That is, each species gets two names, as first proposed by Linnaeus in the 18th century. The first, capitalised, names a genus, a collection of related species. The second is more, ahem, specific.

Linnaeus' system was devised for plants in the first place, many of which had common names, and these were often transformed or alluded to in the new system. Microorganisms were gradually named too, as more and more were identified over the century following his first inclusion of microbes in his system (as 'infusoria') in 1758. Bacteria, being invariably new discoveries, were named pretty much any way that took someone's fancy – the name might be a Latinate version of their shape under the microscope, a reference to where they were first found, the name of their discoverer, or something more whimsical.

Some are marginally informative. A *bacillus* is rod-shaped, a *coccus* spherical, *vibrio* like a comma. So *Vibrio cholerae* is a comma-shaped organism associated with cholera. Hurray.

A few descriptors add a bit more information. Species that cling together in strings of spheres like a necklace are *Streptococci*, but if they cluster like grapes they are *Staphylococci* – both types common enough for most of us to have heard of them.

But that is just the start. There are innumerable more species names. And there are names above this level. A phylum is a collection of genera. So the phylum *Firmicutes* – which have a firm, that is, rigid cuticle or cell wall – includes genera such as *Lactobacillus*, *Clostridium*, *Eubacterium* and *Ruminococcus*, each with its own collection of species. And each species may have many sub-species, or strains. At this point, a kind of nominal exhaustion sets in and a newly characterised variant is usually given a number.

As we are still registering more and more bacterial diversity, there are a *lot* of names. I am going to use them here so that anyone interested can look them up, but it usually does no harm to think of them as bacterium X, Y or Z. Often the names tell you little. It is what the bacteria do, and precisely what genes they have, that matters. And, to be honest, often the names could hardly be dafter. So no points for whoever decided that one of the genera in the phylum *Bacteroidetes* would be called *Bacteroides* (I'm sure there is a history there, but I don't think I care). And a negative score for the namer of a species that we will have occasion to marvel at for its metabolic virtuosity in Chapter 5 – *Bacteroides thetaiotaomicron*. It sounds impressive, until you learn that thetaiotaomicron is a fraternity house at Virginia Polytechnic Institute in the USA.

The professionals learn to navigate this morass of historical palimpsest and small vanities. The rest of us cope as best we can. But these days we have much more precise ways than comparing names for judging whether bacteria are related to each other, and how closely, as well as teasing out which ones live in particular locations and just how they are making their living. An explosion of new techniques has finally allowed a full appreciation of the richness of our own microbial load. In its way, the ability researchers now have to recover information about microbes that was previously hidden, even if the microbes concerned were our intimate companions, is as remarkable as the results it is revealing. So let's look at the tools at our disposal for investigating the microbiome, before delving deeper into what they have uncovered.

3 | Invisible lives

Microscopes showed that bacteria exist. But looking at them, like inspecting my body in the mirror, is informative only up to a point. Many very different bacteria look very much alike. To understand how they live, and what they can do, calls for new ways of looking. Even getting a good idea of what kinds are there involves looking inside – or at least examining material that has been recovered from inside – their cells to inspect the information therein. Science, in a two-step with technology, has got astonishingly good at working out new ways of looking at things to amass more – or different kinds of – information.* These are the keys that have unlocked our microbiome in the last couple of decades.

For decades, there was a nagging problem in microbiology of us. We knew there were many, many microbes, mainly in the gut. But most of them resisted efforts to bring them out into the lab. They were either killed by oxygen, or needed

---

* A pause here for philosophers to think about whether new ways of looking mean we *see something differently*, or see *a different thing*. Daston (2011) offers more food for thought on that – the kind of philosophical discussion I like because it involves looking at what scientists actually do. My own, unphilosophical answer is 'probably a bit of each'.

some elusive chemical to survive. They just would not grow in culture.

By and large, the problem was noted, and then ignored; in human studies as in the rest of microbiology. It was highlighted again in the mid-1980s as the 'great plate count anomaly' by James Staley of the University of Washington.[1] He and a colleague reminded everyone that the number of bacteria you can see under the microscope in a fresh soil sample, for instance, is thousands of times larger than the number from the same sample that you can persuade to grow in a culture dish. They reviewed other methods for assessing microbial diversity that had appeared since the 1930s and encouraged everyone to do better at capturing the full range of microbial types. But their review for microbiologists mainly concerned water and soil ecosystems. For medical microbiologists, there was still plenty of other research concentrating on a relatively small number of pathogens to be getting on with. As long as that feeling persists, science is surprisingly good at the kind of selective attention that allows known discrepancies or anomalies to lie uninvestigated.

No more. Now everyone agrees the human microbiome is complex and important. Settling for study of the easy-to-grow species – basically the microbial equivalent of weeds – just won't do. And ingenious ways are being found to get at the more sensitive. They are what allowed David Relman to add a slew of new species to the inventory of bacteria found in the mouth in his first, simple study.

Seeing the invisible

Investigating the microbiome does not deal with the ultimately small. That is the province of physics, probing particles and interactions so far beyond our senses they boggle the mind. Still, microbiome science depends on making visible things we would never normally know about. Each new look throws up new hypotheses about what has been found, and what it does – hypotheses that call for new kinds of experiment. This has taxed many people's ingenuity since the millennium. Here is a rundown of the microbiome analysts' toolkit, and what they have been able to achieve with it so far.

New worlds first need to be mapped, and that is well under way. In biological terms, this is like the stage in the development of the life sciences when natural history was the order of the day. Naming, labelling, cataloguing organisms and recording their habits and behaviour was about as scientific as things got for quite a few centuries.

At that level, there is *plenty* of information. Information is what modern technology is good at producing and storing. Still, it is hard to grasp just how good life scientists' tech is these days.

The key technology here is DNA sequencing, which reveals the order of the bases – we can regard them as chemical 'letters' – that are strung together in each linear DNA strand, and encode the genetic information it carries. The rate of development here has exceeded any technology, ever. The improvement in performance of computers over the last half-century – usually expressed as the number of processors on a single chip – is often said to follow 'Moore's Law'. Intel's co-founder Gordon Moore suggested in 1965 that the number of transistors in a single integrated circuit on a chip doubled every two years and would go on doing so. So far, his 'law' has proved correct, and the soaring power and shrinking cost of computers is the result.

DNA sequencers are leaving chip designers for dust.

The planners of the Human Genome Project relied on DNA sequencing technology getting faster, but the results exceeded all predictions.

The first complete human genome, with its 3 billion base pairs, took thirteen years to work out, more or less accurately, and the whole vast effort cost almost $4 billion. Manufacturers of commercial DNA sequencers are now marketing machines that will do the job in a day, for $1,000.

The main gains are quite recent. The US National Human Genome Research Institute has plotted DNA sequencing cost against time on a logarithmic graph that compares the trend with Moore's Law. Computer processor cost appears on such a plot as a slowly descending straight line. The DNA line begins to descend faster than the Moore plot in 2007, and falls away faster and faster until now. Result: a raw megabase (1 million bases) of sequence cost a bit under $10,000 in 2001, came down to $1,000 in 2004, and in 2014 cost … around half a cent.

Fast, cheap, reliable sequencing is a transformational technology. Basically, anything that contains a nucleic acid sequence can be mined for information. The amount of DNA sequence information being archived is increasing at an almost incomprehensible rate. A state-of-the-art sequencer that ships ready to work on the lab bench can, under ideal conditions, generate 100 billion base pairs of sequence a day. A 2013 estimate has 15 petabases of new DNA sequence a year – that's a thousand million million DNA letters – being recorded worldwide.[2]

Even with ready access to DNA sequencing, though, there are still trade-offs in any particular study.

Sequencing a single genome is now a routine job, depending mainly on careful preparation of the samples that go into the machine. But the DNA needs to be pure, that is, all from a single organism.

For your primary cells, this isn't a problem. Scrape that finger inside your cheek, and the cells on the end will do at a pinch. The original Human Genome Project took blood samples from volunteers. As the human DNA complement is relatively large, the DNA was extracted, broken down by enzymes into more manageable pieces – 100,000 or so base pairs – which were then 'cloned' or reproduced in tame lab bacteria, whose fast growth meant they provided a large quantity of Bacterial Artificial Chromosome (BAC) DNA as their population blossomed. These BACs were then extracted again for sequencing.

But the smaller genomes of free-living bacteria are harder to get hold of in pure form. Those that will grow in lab cultures can be dealt with easily enough. The rest stay as an impossible mixture, containing who knows how many different organisms, each contributing a tiny quantity of DNA to the total.

There is a way round this, permitting culture-free analysis, combining the growing power of DNA sequencing with improved knowledge of the biology of bacterial life. Pick up life as it comes, wherever you find it. It might be a bucket of seawater, a handful of soil or, for our purposes, a lump of human excrement. Inside, whatever else you find, there will be a mass of bacteria, viruses, and perhaps other, larger organisms. Don't bother to separate them out. Just extract all the DNA, break it into manageable pieces, and sequence the whole lot.

The advent of industrial-scale genomics means scientists now have access to an ever-expanding database of DNA sequences. Once all this information is stored in electronic form, instead of in DNA itself, computers can scan it and compare new DNA samples with the ones already in the bank. They may find the same letters in the same order – a perfect match. Sometimes this is part of a gene whose function – perhaps to guide the synthesis of a particular protein – is already known. Sometimes it will vary from an existing gene in small ways. Sometimes it will be the kind of thing that looks like a gene, but for an unknown product. Stretches of DNA that code for proteins have special signals for where the cell's own machinery for reading DNA should stop and start, for example. With careful interpretation, the 'everything in the bucket' sequencing approach can tell you a lot about the population of the bucket.

It still isn't exactly easy to make sense of what is there in key areas of the human microbiome, though. The human colon is probably the richest and most varied ecosystem on the planet. Separate out all the genes, and you recover plenty that have never been seen before. And because this kind of mass sequencing detaches the gene from the organism, it gives no direct clue to what individual kinds of bacteria, or other things, are there.

Biologists cut through this to get at least a broad picture of how many different species are present by fixing on just one gene from each bacterium. This works because the gene in question is extremely similar in every bacterium, but has small regions that vary much more. In a DNA sequence, that usually denotes a gene whose sequence preserves portions that are needed for an essential function. This one is about as essential as you can get. It is the gene for a piece of DNA's molecular cousin, RNA, that forms a big chunk of the indispensible nanomachine, the ribosome, that takes the genetic message derived from a stretch of DNA and uses it to assemble amino acids into a protein molecule.

One such is the gene for 16S rRNA (named, prosaically, for the speed with which it moves when you put a suspension in a centrifuge, a standard lab technique for separating large bits of cells), which together with some proteins makes up the smaller of the two main sub-units of the ribosome.

Working with this sequence has unbeatable advantages. It homes in on bacteria, since eukaryotic ribosomes are different. It is present in large amounts in the cell, so can be extracted relatively easily. And its 1,500 base span was manageable for sequencing some years ago.

Analysing the genes – just the one, all at once, or species by species?

The 16S rRNA sequence has an important place in the recent history of biology, which helped it become the standard tool for first-level analysis of the microbiome. In fact, early work on 16S rRNA began even before the full sequence was feasible. Carl Woese, who died in 2012, used it to redraw the entire map of life. He began comparing sequences of short fragments of the RNA – oligonucleotides – from different bacteria in the 1970s. Plotting the relationships between microorganisms from the family trees of these sequences led to a major rethinking of the overall structure of life on Earth. Woese found that there was a third kingdom of life, now known as the archaea, quite distinct from the two already known, bacteria and the larger, nucleated-cell eukaryotes. The archaea, like bacteria, are prokaryotes, which were formerly thought to all be related. But they are distinct from bacteria. They were originally seen as rather exotic, inhabiting unusual environments, but we now know that there are archaea everywhere. The human microbiome has its share, too.

Adding an entire third division to the full set of all living organisms rewrote the textbooks, and established 16S rRNA sequencing as a key technique. But Woese's initial work depended on sequencing RNA fragments obtained from pure cultures. More recently 16S analysis has moved on, allowing researchers to pick out the DNA sequences that generate 16S rRNA from wild mixtures of organisms.

While the concept stays the same, it is pretty complex in practice. The samples go through a series of steps: breaking open cells, extracting DNA, then finding all the 16S genes using DNA primers that recognise regions at the start or end of the gene that do not vary at all between species. Then the DNA fragments tagged this way are amplified using a cyclical treatment invented in the 1980s known as the polymerase chain reaction (PCR), which allows minute quantities of DNA to be copied rapidly many times over. Finally, they are sequenced.

The payoff, matching one or more of the variable regions of the 16S gene sequences to ones already known, is these days usually in the hands of computer software linked to DNA databanks that hold tens of thousands of the sequences. Like most molecular genetic routines, what was once a laborious procedure that PhD students cut their teeth on is now often automated. It is possible to make short portions of many such sequences and attach them to a 'microarray', the DNA analyst's equivalent of the microchip. The sequence matches can then be read directly from the sample. What once required a lab to mobilise a set of state-of-the-art skills is consigned to a machine, but the researchers still need to know the exact details of the steps leading up to the results. There are numerous subtleties of method, including how the sample is processed and the DNA is extracted, and exactly which parts of the sequences are compared – the 16S gene has nine of the crucial hypervariable regions and their many differences are not spread evenly between bacterial species. It matters which ones you use. The power of modern DNA methods, especially the ability to amplify tiny amounts of DNA by copying the molecule over and over, also means that they are as sensitive to contaminants as they are to things that were actually present in the original sample. This is a real bugbear in microbiome studies. A careful test in 2014, for example, showed that standard DNA extraction kits many researchers now use are often not sterile, as advertised. If the original sample is from a site with a low microbial population, the results can easily be skewed by bugs introduced by the researchers.[3]

The journal *Nature* commented that the exposure to contamination risks 'adds to concerns in the scientific community that sequencing technology has developed so fast that in some cases it has outpaced scientists' ability to use it'.[4] However, while there is always some fuzziness in the data, 16S readouts tell you things about a sample of uncultured bacteria that could not have been pulled out from a mass of DNA before. And although now very widely used, it is just the first level of analysis. Getting a clearer picture of what a sample contains, beyond a broad population structure, means going deeper into the DNA.

That, too, is now possible. Again, it can be done working with fragments of DNA extracted from a whole sample. In this approach, the system is set up to sequence *all* the bits of DNA you find. Originally, this technique, known as shotgun sequencing, was applied to a single species, with researchers trying to piece together a whole genome sequence from all the pieces, some overlapping, some separate.

Now, the technology is powerful enough for a brute force approach. Never mind how many species are in there. Just break up the DNA, sequence all the fragments, and see what sense you can make of the vast, mixed-up library of sequences that result.

This, then is metagenomics, which yields information about the entire genetic repertoire of a community of organisms, *even if you don't know which organisms are there*. Again, the analysis is often at the limits of what biologists, chemists, and computer specialists working together can do to extract signal from noise. They are working on the equivalent of thousands of jigsaw puzzles all crammed into the same box, then given a good shake, and mostly without much clue what the original pictures looked like.

Metagenomics does offer a powerful way of treating previously unusable samples, though. And the more complete genome sequences that are worked out and filed in the growing reference databanks, the more effective it gets. If a microbiome looks interesting, we can now discover what it contains in as much detail as the budget will allow.

From surveys to experiments

Then comes the part the technology does not make so easy, working out what it all means. That equates, perhaps, to the move from natural history to a deeper scientific understanding of what has been described and catalogued. That calls for a combination of theory and new experiments. Otherwise, metagenomics risks justifying a jibe from Sydney Brenner, one of the founding thinkers of modern molecular genetics, and becoming 'low input, high throughput, zero output biology'.

What will help avoid that? There are plenty of different ways of doing biology, but three general approaches are on offer. One is to figure out how to do controlled experiments on the microbiome. This is pretty hard in people, for obvious reasons, even when the experiment is ethical.* The answer, therefore: turn to other species' microbiomes. They all have them, so there are plenty available for comparison – and the list of those that have been analysed grows longer.

One key to more controlled comparison is lab animals that begin life with no microbiome. They can then be furnished with one tailor-made to answer an experimenter's question. The main model system here relies on germ-free mice, which were first raised in the lab 50 years ago.[5] They are delivered by caesarean section and reared in sterile chambers. The researchers have to do all their work through hermetically sealed sleeves with gloves on the end. It is an expensive and awkward business that had almost fallen into disuse when the new wave of microbiome research started to build. Sarkis Mazmanian of CalTech, a MacArthur 'genius' award holder who we shall meet again in Chapter 7, related how when he began work on how microbes affect the gut, in 2002, he found no one around who knew how to raise germ-free mice. 'I had to persuade a retired research technician to help me set up sterile chambers and teach me the ways of "sanitary engineering". Rather than the old steel and glass contraptions that he had used in his day (50 years ago), we were able to procure nicely modernized chambers with plastic bubbles that held up to four mouse cages. After my first few chamber contaminations, I began to understand why researchers rarely use germ-free animals'.[6]

A decade on and germ-free mouse production lines are

---

* Experiments with a bacterial toxin responsible for some of the symptoms of typhoid were being done on volunteer prisoners in the USA as recently as 1970, but times change ... See Hornick (1970).

established in many more labs. This is good as mice have long been a mainstay of genetics and biochemistry research – and there are standardised strains available from lab suppliers. There are now other germ-free creatures to work with too, including rats and, most recently, zebra fish. The latter oblige scientists by growing transparent embryos that make it much easier to follow developmental events as they unfold.

All these models are compromises. Humans are like mice, or even fish, in some ways, unlike them in others. Any results from these creatures are only pointers as to what may happen in people. Could there be a better model for us? Well, the pig qualifies on several counts. It is a better match for a human than a mouse in terms of its size, its digestive system, its overall metabolism *and* its microbiome. But let's face it, it is never going to end up in as many labs. As one researcher who has made the effort, Colin Hill of University College Cork, cautioned at a conference in 2014, 'a fully grown pig with diarrhoea is no fun for *anyone* involved in the experiment.'[7]

Living animals, and their microbiomes, remain complex, and it is hard to see what is happening inside while they are alive, so there are a range of ways of setting up experiments that are like part of an animal, or a microbiome. Intestinal cells can be cultured and encouraged to grow into something resembling the intestinal lining, for example. Or researchers can try to establish a microbial ecosystem in a flask, or a series of flasks. The shifting population inside the bioreactors can be sampled and measured and the feeding solution and effluent analysed, all to give an idea of the overall functioning of the system.

This helps if your goal is to try to find a mix of microbes that could repopulate the gut of someone whose microbiome has been jolted into a new state by a nasty condition like

Crohn's disease, for example. But it does not satisfy anyone who wants to go down to the level of molecules. Cells are where the biological action is, but it is molecules, large and small, that make them go.

How far down into the molecular detail you want to go depends on what kind of explanation you are after. Experimental set-ups need hypotheses to test, and they depend on ideas – which are the vital complement to all these clever ways of probing the microbiome. Study of the superorganism involves the whole of biology and it pulls in ideas from every part of the subject.

Some of the ideas come from theories about how life developed. As the great theorist Theodosius Dobzhansky said, 'nothing in biology makes sense except in the light of evolution'. Darwin's natural selection, in its modern form, remains biology's 'theory of everything'. But it tends to be better at explaining why things work the way they do, not *how* they work.

For that, the answer depends on what you want to understand, and at what level. The precise details will certainly depend on atoms and molecules in the end. Big effects arise from lots of small things happening. This is the basic assumption of one strong research tradition in biology, which takes things apart piece by piece. This strategy is extraordinarily powerful, and often yields insights of great beauty – the way DNA's entwined pair of helices fit together to provide a reproducible store of information is the best possible advertisement for this approach. The shape of the actual molecule doesn't explain everything about how genes work, but it is a pretty good start.

Much else in biological systems, human and microbial, depends on the shape of molecules, too. The immediate problem, as with those DNA sequences, is that there are so many of them. I agree in principle with the researcher at a microbiome conference who said at dinner that, 'of course we won't really understand what's going on with any of these things until we can explain it in molecular terms'. In practice, though, this does not always seem helpful. It can simply mean generating endless statements of the form 'A interacts with B, which affects C, which triggers a response from D and ... generates result X'. It's impressive when you can work that out but the result still seems as much descriptive as explanatory – although if it allows you to block nasty result X by interfering with A, B, C, or D it may be enough to be going on with.

I don't want to remember thousands of formulations like that, and it will take a *very* long time to generate them in many cases of interest. So we will need some higher-level concepts to make sense of the causes and effects operating when we and our microbiomes interact. With this in mind, let us move on to investigating some of the torrent of results the researchers are generating with all these new techniques.

4 | Microbes, microbes, everywhere

The total area of the average person's skin is a couple of square metres. Lay down a typical bacterium in that space and it has roughly as much room, proportionally, as a lone person who lies down on a patch of the Earth's surface the size of all 50 states of the USA.* On a more everyday scale, lay bacteria down close together and it would take a million to cover the head of a pin.

This difference in scale makes many surveys of our microbiota very broad. If our skin, for example, looks like a continental landmass to bacteria, we will want to know something about the geographical distribution as well as the total microbial population. And the body's inner surfaces, as we'll see, are a lot bigger than the skin, so there the problem is magnified.

Researchers can now sample from as many sites as people are willing to offer – the record is probably a study of skin that looked at 400 different places on each of a pair of heroically

---

* This is a very rough estimate, but I reckon correct to an order of magnitude. I am assuming a human lying down occupies three square metres, while a bacterium a micron in length would take up a million millionth of a square metre. The land area of the USA is around 9 million square kilometres.

patient subjects – and get a pretty accurate microbial roll call. New results in science invite new questions, so each fresh survey has been followed by more sampling, more analysis, more detail. But getting the measure of the superorganism first demands a look at answers to the most basic question: who's there?

The most basic answer is, not everyone. That simple finding is important. True, scientists are newly aware of a startling number and diversity of microbial species in and on your body. But the diversity is still limited. It is a reminder of a famous saying in microbiology, coined by the Dutchman Baas Becking: 'everything is everywhere, but the environment selects'. All bacteria have the potential to be ubiquitous, but where they are found depends on the precise environmental conditions – including the presence of other bacteria.

Suppose we put together every study of the human microbiome, ever, and pool all the bacteria found. Estimates of the number of bacterial phyla (the taxonomist's main category that holds a big collection of species) on Earth range between 50 and 100. In other words we do not really know how many phyla there are. But we do know that only a small number contribute to our microbiome. Very small colonies of others may yet turn up, but it looks like no more than a dozen phyla have got a purchase on people. Just half a dozen are dominant, accounting for 99.9 per cent of the total.

We know bacteria reproduce fast, and can quickly fill any niche that allows them to grow. So the absence of the majority tells us that they are not getting aboard the good ship *Homo sapiens* by any random process. There is selection and control going on here. You expect that in a system where co-evolution has been going on for millions of years.

That is reinforced by the next general finding. Not all the microbes are found in every body site. The environments we offer are very different, and so are their bacterial populations. To get a well-rounded appreciation of the superorganism calls for zooming in to look at these sites in more detail. But it is useful to first get a view of the world map of the human microbiome, and a pioneering US collaborative project to provide just that is a good place to start.

The widescreen view

The US effort was christened the Human Microbiome Project (HMP), to encourage researchers – and funders – to see it as a natural successor to the successful Human Genome Project. Launched by the US National Institutes of Health (NIH) in late 2007, it was intended as a baseline survey of healthy people. A consortium of labs worked on samples from 242 adults, taken from fifteen body sites in men and eighteen in women. (Most of the men and women came from two cities, Houston and St Louis.) The total number of analyses reached over 11,000, with each donor asked to give repeat samples twice over two years. Around half the samples yielded the basic 16S rRNA sequences that permit identification of microbial types without showing the precise strains present or their detailed genetic make-up. A small set, 681 samples, were sequenced after the DNA had been chopped into small bits, and the sequences matched with known genes and parts of genes, giving an idea of which proteins the microbes could make. Finally, 800 bacterial strains isolated from these 242 people had their genomes sequenced completely. The whole lot cost a lot less than the Human Genome Project, but the total spend was still $170 million.

The advance of technology meant that this sum paid for

more data than biologists had ever seen. The consortium of labs working on the project ended up dealing with 18 terabytes of data, some 5,000 times as much as the Human Genome Project dealt with. Their first main analyses came out with appropriate fanfare in the most-read journals, *Nature* and *Science*, in mid-2012.

They confirmed the other main finding about the human microbiome: how complex and variable it is. The widescreen image of the microbiome of healthy Americans from the Human Microbiome Project was wide indeed. Between them, the people who took part carried more than 10,000 microbial species, which provided their several ecosystems with 8 million genes, or 360 times more than their human hosts keep in their own cells. The sheer number of different microbes vying for living space on each of us means that there is every chance the results for you or me will vary from our neighbour's. When it comes to the microbiome, 'Even healthy individuals differ remarkably', the consortium declared at the beginning of their main summary paper.[1] As Rob Knight of the University of Colorado explained, this meant that for any microbe that one person has in large amounts, there will be someone else with hardly any of them.

The next headline finding was more of that diversity. Individual body sites differ. A lot. But it was the fact that microbiome samples from the same body site taken from different people vary widely that was more unexpected – and unwelcome to the promoters of the project. The Human Microbiome Project was sold to funders as like the Human Genome Project. It would, research strategists hoped, reveal a reference microbiome, just as the earlier effort established a consensus human genome sequence. That would be a launch pad for further work to probe the consequences of small differences between people. But now we can see that there *is* no reference microbiome: just many, many microbiomes. The authors of one paper departed from the normally restrained language of science to describe the finding that 'no single phylum was present at any site in all individuals' as 'staggering'.[2]

You can, of course, smoosh all the data together and summarise the collective microbiome of the people in any given study, or a whole bunch of studies. So that is what they did.

These twenty dozen healthy US adults, with their 10,000 microbial species* between them, donated their microbes from a collection of sites chosen partly for interest, partly convenience. Nine of them were in and around the mouth – covering saliva, two sorts of plaque, and the tongue, roof of the mouth, inside of the cheek, gums, tonsils and throat. There were four skin sites: the left and right inner elbows and the crease behind the ears. On the other hand, the relatively inaccessible lower gastrointestinal tract, with its much larger microbial load, was simply represented by what it expels, the faeces. Microbes from these sites were taken from everyone who joined the project. Women gave three extra samples, from the entrance, middle and back of the vagina.

The results confirmed the main contributors to our common bacterial load. Most of the bacterial portion of our microbiomes are from just four phyla: *Bacteroidetes*, *Firmicutes*, *Actinobacteria* and *Proteobacteria*. The first two predominate in the average healthy gut, especially in the colon, although *Firmicutes* are

---

* The definition of species here is purely technical, relating to 16S rRNA sequences. They are known formally as 'operational taxonomic units' or OTUs. It is OK to treat them as species when trying to get a grasp of microbial diversity.

also found pretty well everywhere else. *Actinobacteria* and *Proteobacteria*, on the other hand, abound in the mouth, and are also found on the skin.

So far, so uninformative. Each of these phyla contains a large assembly of species. The charmingly named *Firmicutes*, for example, are just bacteria that have a relatively strong outer wall. That single trait is used to distinguish them from the wimpy *Mollicutes* (*molli* being Latin for soft), in which it turns out I have a personal interest (of which more in Chapter 5), which hardly have a cell wall at all.

That makes results from first-level microbiome studies hard to read for significance. Try to interpret a change in the gut microbiome when all you know is the shift in the ratio of two bacterial phyla, says Harvard epidemiologist William Hanage, and you are as well-informed as someone who thinks an aviary containing 100 birds and 25 snails is identical to an aquarium with eight fish and two squid. Both have four times as many vertebrates as molluscs.[3]

As soon as you shift below phylum level, though, the amount of diversity can be a bit bewildering. The Human Microbiome Project inventory ended up with 1,000 bacterial species in the mouth alone, 440 inside the elbows, and 1,250 behind the ears. The vagina has the simplest microbial population of those sampled, often harbouring just a few species of the genus *Lactobacillus*.

Elsewhere, there is invariably a long – often very long – list of species, and it differs over quite small regions. Mouths have different population mixes at each of the sites sampled. Skin has different microbial flora in dry, moist and oily areas (you know where they are), and in small structures like hair follicles and sweat glands. The gut, the largest microbial vessel of all, has many different niches for microbes but this overview was not geared up to reflect that. One scientist cautioned that relying solely on faecal samples to understand the gut microbiota is like trying to visualise a Ferrari's paint job by sniffing the gas from the exhaust pipe.

So many, many species! It would be possible, but pointless to list them all. Some of the names might be familiar, most less so. But what we really want to know is what they all do, and how that differs: between body sites, between people, and in health and disease. Getting into that calls for a closer look at our various microbial niches, and then for overlaying survey maps with other kinds of evidence. Let's take a look at each of our member ecosystems in turn, summarising its main features. The most complex of all, the gut, deserves a chapter to itself – for which see Chapter 5. Here, I will look at the range of other sites – which are less complex, though not necessarily simple – starting with the place that microbes, like other things, encounter first: our skin.

Skin-deep

Skin, in constant contact with bacteria, is also the microbial home that is most obviously a varied landscape. The persistent moisture between the toes that encourages fungus is a noticeably different environment from the tips of your fingers, your armpit or groin, or the large expanses of your back.

Some of the microbes that graze on skin – which is constantly sloughing off dead cells from its outer layer that can provide food – draw attention to themselves by making life uncomfortable, or just a bit smelly. This, plus ease of sampling, means the skin microbiota have been studied for many years. As Jessica Snyder Sachs points out in her 2007 book *Good Germs,*

Bad Germs, a volume with the contemporary-sounding title *Ecology of Human Skin* appeared as long ago as 1965.

As with other regions, that does not mean we know all about it. For example, *Staphylococcus epidermis*, a bacterium named after its favourite habitat, is an almost invariable coloniser. It is one of several species long thought to crowd out less innocuous competitors – in particular *Staphylococcus aureus*, which is a frequent cause of skin infections and worse, though it can also live quietly in the nose and do no harm at all.

However, in 2012 careful experiments in germ-free mice showed that *S. epidermis* also helps activate immune responses in the skin that are needed to ward off pathogens. Specifically, the presence of this bacterium allowed the mice to escape the effects of *Leishmania major*, a completely unrelated eukaryote that in people causes the tropical disease Leishmaniasis.[4]

Skin's varied microbial population also demonstrates that we cannot avoid bacteria. We just influence which ones will take up residence. Unlike the gut, skin's job is relatively simple, keeping the outside out and the inside in. It keeps bacteria down using two types of sweat glands – sweat does not need to be so salty just to cool you down: the high salt content helps retard bacterial growth. Fatty exudations from sebaceous glands offer an additional microbial shield.

That does not mean there are no microbes in areas where these glands are most active. It does mean that the ones that live there can tolerate a salty, acidic milieu, or can eat sebaceous secretions. The notorious *Propionibacterium acnes* finds the low oxygen levels of the sebaceous glands' pits congenial, and has a battery of enzymes that allow it to eat the lipids it finds there. As hormones step up the glands' output during puberty, *Propionibacterium* species including this one increase, and the condition it is named after often follows. There is a trade-off, though. The bacterium also produces fatty acids from the lipids it digests that have antibacterial effects. Unfortunately for acne-ridden teenagers, they work only on other species.

Along with these major players, the skin supports a host of other microbes, as well as fungi and mites, the latter not quite 'micro' enough to make it into this book. They vary a lot between people, and over time. The skin microbiota may be more diverse and variable than anywhere else in the body. This makes intuitive sense given its contact with the outside world. Each of us carries an epidermal menagerie, as confirmed by studies like one that demonstrated you can trace the pattern of one person's fingers on a computer keyboard by identifying the distinct microbial populations transferred from fingers to keys.[5]

A comprehensive review of skin studies in 2013 described the diversity of skin microbiota in healthy human adults as 'staggering'.[6] In one study, samples from six people's forearms had fewer than 10 per cent of bacterial genera in common. The same person's two hands have different microbial communities, overlapping by only 17 per cent. We do, however, tend to share more skin bacteria with our partners, and our pets, than with strangers. The same authors suggest that all these results 'challenge our concept of the human microbiome'. Perhaps, they say, the thing to discuss is the microbiome of 'all the residents of our home and place of work'.

Another study underlines that as well as supporting microbes, the skin sheds them constantly, and at such a rate that any room we are in quickly acquires our own microbial signature. We move in a cloud of microbes that we carry with us. The US-based Home Microbiome Project published first results from a long-term observation of seven families and their homes in 2014. Their sampling took in sites all around the home, as well as each family member, and they found strong links between the people and their surroundings. Floors are microbially more like feet than walls, while light switches pick them up from hands. People are a richer source of microbes than the rooms they live in, so the traffic is mostly one-way. One couple in the study moved – from a hotel room to a new home. Their new surroundings rapidly assumed the same microbial profile as the one they had imprinted on the room they had recently left.[7]

Another striking demonstration of diversity is one that gets away from the limitations of the majority of early microbiome studies, which sample people in the affluent North of the globe. A Yale University-led comparison of women in Tanzania and in the US showed big differences in the microbes found on their hands.[8] The groups were small, and the 29 Tanzanian women were child-carers while the fifteen US citizens were not (the researchers went for convenience there and sampled graduate students). Still, the differences were easy to detect. All the common bacteria were there in both groups, but hands in the USA yielded a lot more *Propionibacteria* and *Staphylococci*, while the Tanzanians had a lot more bacteria often found in soil. The Tanzanians also had more than ten times as many bacteria per square centimetre in total, but from only two-thirds as many species. As the Americans spent most of their time indoors while the Tanzanian women lived in open-air dwellings, this is not too surprising, but it does reinforce the idea that the skin microbiome is strongly influenced by everyday contact with our surroundings. It also, perhaps, cuts across assumptions that being outdoors goes with microbial diversity.

Although the skin encounters plenty of microbes

serendipitously, like other parts of the body it has also evolved to accommodate some species as long-term cohabitants. We tend to think of skin as easily cleansed – washing our hands as we were taught. That is a good thing to do, but to dislodge recently deposited bacteria rather than eliminating bacteria altogether. The latter, even if it were possible, would merely lay the skin open to colonisation by less desirable species.

What is more, there is intriguing recent evidence that some bacteria penetrate the lower layers of our skin, beneath the epidermis. Richard Gallo's group at the University of California went looking for bacteria in the layer beneath, the dermis, and in fatty tissue beneath that.[9] They found distinct bacterial communities in both compartments. That slightly disconcerting finding is significant on two counts. It gives bacteria that have found their way into the lower reaches of our skin more direct contact with our immune system, which has complex interactions with the microbiome that we shall take up in Chapter 7. More generally, it is an addition to a growing list of body sites we used to think of as sterile that turn out to harbour at least some organisms that originate elsewhere. We are more thoroughly inhabited than we knew.

The mouth – many microbiomes in one

The mouth, where observation of our own microbiome began all those years ago, continues to surprise, as David Relman showed with his pioneering DNA analysis. It continues to feature in some of the simplest investigations of our microorganisms, and in some of the most sophisticated.

If we want to start simple, what could be simpler than a kiss? Yet, from a microscopic perspective, kissing is pretty complex. A 2014 study, dubbed the 'Makeout Microbiome' paper by one observer, showed how much microbial transfer happens when kissing progresses from a peck on the cheek to a more intimate exchange. Remco Kort's group in the Netherlands asked 21 couples about how, and how often, they kissed, and got kissers to help with an experiment in which one partner drank probiotic yoghurt between kisses. They confirmed what you would think: that kissing was a pretty good way to shift bacteria from the yoghurt-soused tongue to the yoghurt-drinker's partner. A ten-second kiss transfers perhaps 10 million bacteria, they reported.[10]

But the long-term results of kissing, like human relationships, are more complex. The surface of the tongue does have a microbial population that is more similar between partners than people chosen at random, but it does not link that strongly with how often they kiss. The immediate transfer leads to established populations of some bacteria, but not others, in a selection influenced, it appears, by lots of other factors. Kissing has an impact, but the ecosystem has the last word.

At the more technically taxing end of the research spectrum, the oral microbiome had a starring role in another landmark paper, appearing in mid-2014, which made new use of an old survey. It's an increasingly common way to make new discoveries in biology. The DNA sequences from earlier research are archived in databases anyone can use, and the computer-savvy can pose new questions whose answers can be teased out from the same mass of information.

Murat Eren of the Marine Biological Laboratory at Woods Hole Massachusetts (a state-of-the-art lab that does much more than marine biology nowadays) asked whether reanalysis of Human Microbiome Project data could improve on the inventory of bacterial types derived from their sample collection.[11]

The test case was the section of the sequence bank with microbes from American mouths; the answer, an emphatic yes.

There were still more bacteria hidden in there because the basic 16S rRNA analysis, while incredibly useful, is limited. In particular, when it looks for distinct bacterial types by comparing some of the regions of the gene that are hyper-variable, it chooses only some regions, and sets a threshold for what counts as a different species (or operational taxonomic unit in the jargon). Sequences that differ in this respect by less than 3 per cent are treated as the same.

This helps avoid overestimating diversity because of small errors in sequencing. But it also results in lumping together some microorganisms that are different in important ways. Eren revisited the same data using a more powerful approach that looks for 'the most information rich nucleotide positions' in a collection of 16S rRNA. The reasoning goes deep into the theory of information and entropy laid down by Claude Shannon in 1948 – but we can treat the method as going fishing in a sequence pool with a finer mesh net.

So fine, in fact, that the new bacterial types it defines – now known as oligotypes rather than operational taxonomic units – may differ in their sequence in this particular gene by as little as a single base pair of DNA out of 1,500.

Eren's team chose the oral microbiome to try out their method because intense study has resulted in a mass of systematically organised information. It is one of the most thoroughly studied subsets of our passenger species. The full richness of the life in our mouths is now open to inspection in the Human Oral Microbiome Database, which in 2014 catalogued 688 species (in the sense defined by 16S rRNA sequences). An impressive 440 of them have been cultured, a larger proportion than scientists have usually managed because the mouth has fewer anaerobic bacteria – which cannot flourish in the presence of oxygen – than, say, the gut. We even have full genome sequences on file for no fewer than 347 of them. This oldest site for investigating our microbes has been pretty thoroughly studied, then. But more ingenious ways of sifting the data show that there is still much to learn.

The new analysis had two main stages. The original Human Microbiome Project data, a collection of 10 million 'reads' of two separate short regions of the 16S gene, was turned into a new classification of oligotypes – 490 from one region and 360 from the other. Then the oligotypes were compared with full reference sequences in the more comprehensive Oral Human Microbiome Database.

Were the oligotypes the same as species? There is no simple answer, just as there is no simple way to say what 'species' means down among the microbes. What sense could be made of all this data, then? Well, (deep breath) ... some oligotypes, about 15 per cent, turned out to be indistinguishable from single species in the catalogue. Some groups of two or more species were indistinguishable by oligotype. And, the payoff, more than 150 species in the database mapped onto *multiple* oligotypes, raising at least the possibility that the original species classification concealed significant differences. Another 86 oligotypes seemed to be distinct from any existing entry in the Oral Microbiome Database.

This holds a general lesson for assessing microbiome surveys. What you find depends on exactly how you look. There are probably hidden layers of diversity awaiting detection. Exploring the microbiome by applying algorithms that process vast amounts of sequence data is a bit like spotting creatures in the jungle with binoculars fitted with lenses that only transmit light wavelengths over a single colour band in the spectrum. Switch to lenses that magnify a different colour and a different but overlapping set of jungle denizens will come into view.

There is more to take from this analysis. The best reason for believing that the diversity disclosed by oligotyping matters is that different oligotypes of what look like very similar microbes live in different places. This is telling you something new about the ecosystem, though at the moment it is anybody's guess what that is. Go through more than 200 people represented in the Human Microbiome Project, and one oligotype might consistently be recovered from tongue samples, another from the teeth. Sometimes bacteria that lived mainly at different sites differed by only a couple of nucleotides in the 16S gene.

The way Eren puts this is charmingly candid about the fact that these are data points whose biological significance has yet to be teased out.[12] He highlights two organisms that differ in their distribution. One is the already identified species *F. periodonticum*, which is important because it is linked with inflammatory gum disease. The other is similar enough to be given the same label, almost, but with an addition. It is typed as '*F. periodonticum* 98.8%'. That makeshift naming, says Eren, is a way to say 'the closest thing we found in the database for this oligotype was *F. periodonticum*, but this is clearly something else'.

The assumption is that such small differences in 16S rRNA go with larger differences in other genes, and hence in how and where the bacteria can live. Confirming that requires a closer look at individual microbes. As there are now more of them to look at, the overall effect of the improved information for now is to make the oral microbiome look even more complicated.

It already scored high on diversity. All twelve microbial

phyla (eleven broad kinds of bacteria and one of the archaea) that have been detected in people are found in the mouth, covering hundreds, very possibly thousands, of species. The largest number of different kinds are found in and around the teeth and the gum crevices, but there are distinct communities at the other sites in the mouth: the throat and tonsils, and on the tongue.

Closer study has shown key features of our interaction with some of the important species, especially on our teeth. Some of them, mainly *Streptococcus* species, bind specifically to proteins or carbohydrates found in the thin layer of mixed saliva and fluid from the gum crevices that coats your teeth. Once they have got a foothold, they themselves offer binding sites for second and third stage colonisers. The result is a complex biofilm of interacting bacteria that covers the tooth surface. There is a large industry devoted to removing it, and even to eliminating oral bacteria, as a way to prevent tooth decay. There are many tenuous connections proposed between one or other portion of the human microbiome and diseases. This one, though, is as clear as it could be, in that we know which bacteria in a biofilm actually corrode tooth enamel by making acid from sugar. Still, it will be a good sign of the benefits of microbiome science if better understanding of the ecology of these communities allows us to develop subtler methods to preserve our teeth than mouthwash or antibacterial toothpaste. I look at what some of them might be in Chapter 11.

An ecosystem for defence – the vaginal microbiome

We are learning that every microbiome will have its own story. It may be relevant for only half the population, but the vaginal microbiome – one of our simplest in terms of community structure – is a great example of how the new tools for tracking microbes test ideas about systems that were thought to be well understood. A long-established, and rather simple, story about one type of bacteria maintaining one useful condition, acidity, has now given way to one of many as-yet-unfinished tales about shifting equilibria in an ecosystem.

The vagina has a long history in microbiology, partly because of intensive study during a long controversy about the causes of childbed fever in the 19th century. A late footnote to that work came in 1891, when a German gynaecologist told a medical congress that he had cultured a new, rod-shaped organism from vaginal swabs.

In the late 1920s, the organism he found was classified as *Lactobacillus acidophilus*. As with many microbial species, it is now known to be a collection of closely related organisms. All have in common that they produce lactic acid by fermenting carbohydrates, including those found in mucus. The conventional wisdom for the ensuing decades was that a hefty population of *Lactobacillus* made for a healthy vagina, by maintaining acidity (measured as a low pH value). That prevents growth of other bacteria that cause disease.

This made sense as a bacterial supplement to unassisted vaginal chemistry. Cells in the vagina's internal epidermal layer, around 20 square centimetres of tissue on average, also get their energy by turning sugary glucose to acidic lactate, and this internally produced acid diffuses across the epithelium into the mucus above. Lactobacilli then lower the pH still further. The vaginal cells and their bacterial guests work together to deter colonisation by other species.

But recent microbiome studies complicate the picture. *Lactobacillus* species are still the most common in vaginal samples – present in around four-fifths of 400 North American women surveyed by Larry Forney's group at the University of Idaho in 2010. But that study[13] also found plenty of perfectly healthy women maintaining their vaginal pH without help from *Lactobacillus*. Similar results have been reported from a study in Japan, and a follow-up study in the US found a quarter of adult women to have few or no *Lactobacilli*. They seem to have recruited other species from the 280-odd found in total to do the same job. But there is a more significant rethink that has had a boost from recent microbiome studies.

The common medical condition bacterial vaginosis has one of those names which is simply a translation of regular words into a more formal-sounding phrase – it means a bacterial infection of the vagina. That aligns a diagnostic tag with the traditional way of understanding infectious diseases, as a result of invasion by a pathogenic organism that should not be there. It is a troubling condition, and associated with a range of serious problems in pregnancy including premature delivery and miscarriage. A risk indicator for pregnant women or, better, a way of preventing vaginosis would have huge benefits.

The variations now apparent in the microbiomes of women who are symptom-free and seem perfectly healthy make this harder to deliver. Trickier still, the variation varies. In the North American population it differs between ethnic groups. The picture is a mosaic. Surveys show five broad types of vaginal microbial community, four of which have a large portion of particular *Lactobacilli*. All five occur in different proportions across four ethnic groups. The persistent patchiness of our knowledge of microorganisms also emerges from this study. The most common *Lactobacillus* of all, *L. iners*, is found in at least some quantity in 66 per cent of all the women studied. It was only identified in 1999, more than a century after the first reports on normal vaginal microbes, as it is choosier than others about which laboratory media it will grow on.

The sums still show that *Lactobacillus* strains are the major contributor to the vaginal population in 90 per cent of white women, and 80 per cent of those classified ethnically as Asian, but only 60 per cent of black and Hispanic women show predominance of this kind of bacterium. Possibly because of this, they also have a slightly higher (that is, less acidic) vaginal pH on average.

However, there is also evidence that the precise strain of *Lactobacillus* matters, and not because of its effect on acidity. Some *Lactobacilli* produce appreciable amounts of hydrogen peroxide, which also discourages other bacteria, and these variants are more effective at preventing vaginosis. That is an odd finding as oxygen levels in the vagina are low, which would make peroxide production less likely, so perhaps they do something else too. Some strains also make antibiotic compounds, for instance. Indeed, a trawl of human microbiome genes for groups of enzymes that synthesise small molecules, and analysis of the products they make, has already uncovered one antibiotic new to science – dubbed lactocillin – that is made by common vaginal bacteria. There may well be more.[14]

Another exploratory study indicated that the vaginal microbial population can also change quite rapidly, with the changes possibly linked to whether the sample donors – who all remained healthy – were having sex, and the original make-up of the microbiome. All this makes any simple diagnostics harder to envisage.

Reading any signs of disturbance from an examination of

vaginal microbes is also complicated by each woman's monthly cycle. Studies that track vaginal microbiota and the menstrual cycle find that there are quite large changes in bacterial populations, but what they are and when they happen may be different for each person studied.

Amid all this complexity, there does still seem to be a connection between *Lactobacillus* and successful pregnancy, though, and we may have adapted to exploit it. Pre-adolescent and post-menopausal women are likely to have little or no vaginal *Lactobacillus*, as changes in oestrogen levels during the reproductive years make this particular niche more accommodating to this kind of bacteria. And the old correlation is still visible in the premature birth rates of black and Hispanic American women, which are higher.

That seems to confirm that low *Lactobacillus* indicates high risk in some respects. Frustratingly for doctors looking for diagnostic clues, though, it looks as if it has to go down as a useful indicator, nothing more. At this particular site, intensive analyses of multiple microbiome samples counts against simple definitions of departure from a normal population mix. It is no longer clear-cut what 'normal' is. One of the standard diagnostic tests for vaginosis uses absence of *Lactobacilli* as a criterion. This is clearly wrong, says Forney's co-worker Roxana Hickey, and based on faulty logic. Their presence may denote health, but their absence does not denote ill-health. Some other species may have taken over their job. In this microbiome, *Lactobacilli* may be sufficient for health, but are not necessary.

The best conclusion for now is that the vaginal microbiome is best considered as an ecosystem, and its behaviour investigated in ways that test out ideas about how ecosystems in general maintain their stability or may be disturbed. Researchers

looking at every individually defined microbiome in our set
have come to pretty much the same conclusion.

Specialised skin – the penis

The penis is somewhat less interesting than the vagina, microbiologically speaking – as a microbial habitat it is mainly just
more skin. The planners of the Human Microbiome Project
presumably thought so. They found the vagina important
enough to sample it at three different sites, but ignored the
penis altogether.

A few studies focused specifically on a boy's favourite body
part have been done, however. They found distinct microbial
communities on the surface of the penis, sampling the area
under the foreskin (if there was one), and in urine samples. The
former were indeed similar to those found elsewhere on the
skin, but they also included the kinds of bacteria often found
in the vagina, including *Lactobacillus*.

A small study of teenagers looked at whether their sex lives
made any difference, and found the answer was 'not much'.
However, there is obviously scope for more investigation of the
intimate microbial exchanges that are bound to happen during
vaginal, anal or oral sex than can be accommodated in a sample
of a dozen adolescents aged between fourteen and seventeen
who were having sex, and another half a dozen who weren't.

If anything, this study made the picture more confusing.
It was already known that the bacteria thought to be characteristic of the vaginal population are found on occasion in
the urethra of males. The obvious assumption is that they are
harmlessly transmitted from a woman to her sexual partner. But
this study suggested that some of them can take up residence
in males before a youth has had sex. The authors caution that

they only did 16S rRNA sequencing, so they cannot vouch for the identity of the species involved, but they ponder whether there is some ecological function for *Lactobacillus* in men as well as women.

This project was also a nice study in the arts of persuasion. The methods section reads as follows: 'Urine was obtained by first-catch void into a sterile collection cup. CS [coronal sulcus – the super-sensitive region under the glans] samples were obtained following training on a flaccid penis model. Participants were instructed to retract the foreskin (if present), and firmly trace the groove of the coronal sulcus circumferentially (using a flocked 4mm flexible handle elution swab). Samples were immediately placed in a cooler, and transported to the laboratory within four hours where they were stored at −80°C until DNA extraction.'[15] Each participant did this not once, but four times: a helpful lot, the teenagers in this study.

Other studies have thrown light on another easy-to-identify difference, between circumcised and uncircumcised men. This has been well researched because of a long-established link between circumcision and a reduced risk from HIV.

The most extensive penis microbiome study to date used before-and-after sampling from 79 men in Uganda who agreed to be circumcised, and compared the results with a similar number who remained uncircumcised until the study was over.[16] The samples were collected during a large trial focusing on AIDS prevention, with the microbiome analysis added on. The big change was a shift from anaerobic bacteria – able to live without oxygen – in favour of strains that are aerobic and flourish in the fresh air that surrounds the unadorned penis where it was formerly shrouded by the foreskin. The marked shift in population was compared by one of the study authors to rolling back a rock and seeing the ecosystem beneath change.

There is something odd about this trial, though, as both circumcised and control group came up with many fewer bacteria a year after the research began. You'd guess that being recruited into an experiment that studies your very own penis by researchers equipped with swabs and vials can alter your general attitude to genital hygiene. The authors do not comment on this in their paper so it remains a guess. It does make you wonder how reliable and repeatable the microbial findings might be, though.

In addition, how circumcision relates to any reduction in HIV risk is not obvious, although there is also a reduction in overall bacterial diversity. The researchers speculate that bacteria living under the foreskin may provoke a mild inflammatory response that results in HIV getting readier access to the class of immune system cells it infects.

All this work also confirms that there are many fewer bacteria on the penis than are found at other, more richly populated sites. However, they do sometimes include species that cause vaginal infections, so the sharing of genital microbes that happens during sex without a condom will come under closer scrutiny in future research, availability of volunteers permitting. For now, the admirably concise headline one website came up with for a report on the Ugandan study seems a fair summary of what we know: 'The critters change after you're cut'.

Microbes where you least expect

That completes the list of minor microbiomes that were recognised before the new DNA era but where a poor image of what is there has become somewhat sharper under the gaze of the new technologies. But those technologies have done more. The ability to recover microbial traces from populations that are sparse, or hard to culture, has revealed life in pockets of our bodies everyone believed were sterile. They include the breast and even the placenta – which I'll discuss in Chapter 6. But to finish this first inventory, here are the recent revelations about two tissues with their own microbiomes: lungs and eyes.

It is pretty obvious that your lungs are exposed to the outside world. You can feel the air passing through your nostrils as you breathe in, then out again. Even so, medical textbooks still say that healthy lungs are sterile.

And the microanatomy and biochemistry of the lungs and airways does look well designed to confine the microbiota to the upper respiratory tract – the parts above the larynx – which are inevitably exposed to a constant stream of incomers. The air we breathe typically contains anything from 10,000 to 1 million bacteria per cubic metre.

The larnyx closes reflexively to prevent choking when it senses a chunk of foreign matter. There are more elaborate barriers and disposal mechanisms aimed at microorganisms. It looks as if, unlike the intestines, the lungs have no use for bacteria so every effort is made to keep them out. Mucus, which becomes phlegm, traps incoming matter, whether bacteria, dust or pollen, and beating cilia sweep it up out of the lungs and into the throat, where it is spat out or swallowed.

This mechanical sweep operates alongside a combination of antibacterial agents, like the enzyme lysozyme, found in large quantities in mucus and saliva, and immune cells that also work to keep the organisms in the air, as well as the ones living in your mouth and nose, out of the delicate maze in the ever-branching interior of the lungs. And so, doctors thought, this normally allows exchange of oxygen and carbon dioxide in the capillaries supplying blood to the finest airways deep in the lungs unsullied by any microbial interference.

If microbes did colonise the lungs, that invariably signalled ill-health: usually bronchitis (simply, viral or bacterial infection of the bronchi, the larger branches of the lungs' gas tubing) and – the diagnosis inscribed on many a death certificate in a geriatric home – pneumonia. In the latter, similar to bronchitis but frequently more serious, a bacterial or viral infection provokes an inflammatory response. If this fails to clear the infection, the alveoli at the end of the last, tiny branches of the lungs clog with dead cells and fluid, and oxygen exchange with the blood is choked off.

However, like many other bits of microbial lore, contemporary microbiome studies have exploded this one. The respiratory tract has its own characteristic population. At least we think it does. It isn't as easy to study as some other regions. The upper tract (mouth, nose and throat) which has its own flora, is easier to sample than the lower (the lungs themselves). But the idea that the lungs shouldn't harbour bacteria was so well entrenched that the NIH Human Microbiome Project's first mapping of species didn't even include them.

Those trying to remedy this omission have not had it easy. Sampling calls for bronchoscopy, in which a tube is passed down the nose or throat. People don't volunteer for a second go, on the whole, so most studies report results from a few people, each of them sampled just once. The samples you do get may be mixed up with species from the nose and mouth, from gastric juices, even from biofilms that coat the inside of the bendy plastic endotracheal tubes used to access the lung. The surefire way to avoid this is to sample from lungs which have been removed during surgery for transplantation, but then you only get to use the diseased ones as the items from the healthy donor are needed urgently elsewhere. This isn't the kind of study that is going to be repeated regularly, either, though one such did show that bacterial populations are different in different regions of the lung. Other less drastic techniques can be used, but they still involve washing samples out of the lungs with fluids or insertion of tiny brushes into the airways. All of this means that most studies so far have relied on samples from patients who have a disease, ranging from cystic fibrosis, in which lung infection is a permanent hazard, to HIV, to the chronic obstructive pulmonary disease which afflicts many smokers.

So far, though, it seems reliably established that healthy lungs do indeed usually play host to some bacterial species – the kind that cannot be easily cultured but can be identified by 16S rRNA sequencing or other recent techniques. But the idea that microbes are invaders bringing sickness dies hard. The finding that healthy lungs have resident bacteria was first answered by suggestions that the results were due to contamination from the admittedly bacteria-laden mouth and nose. However, the consistency and repeatability of the findings in the most carefully conducted studies leave little doubt. The normal lung microbiota are relatively sparse, and simple, but the lungs are not bacteria-free.

That relative simplicity has encouraged some researchers to think through the advantages of viewing each microbiome in different parts of the body as an ecosystem. Ecosystem concepts are often proposed as one key to understanding the microbiome, but it is not always clear how to apply them.[17] It is all very well saying that the trillion-fold microbial swarms in the colon are an ecosystem but it is such a complex one that ecosystem thinking may not help that much in analysing what is going on.

Robert Dickson of the University of Michigan Medical School suggests three ecologically inspired ways of approaching the lung microbiome. Think of the lungs as a series of possible colonisation sites for microbial drifters, he suggests. The best model for how this works out is the theory of island biogeography, developed in the 1960s to account for the populations of islands far out in the ocean. The simplest case is a new island left after a volcanic eruption. It starts lifeless, but gradually acquires colonists that drift, swim, fly or even get carried on rafts of fallen timber. The number of species that encounter the island depends on how big it is, and how far from the nearest landmass with a stable ecosystem. The number that survive also depends on size, and on other local factors. The models that were developed to capture all this can be applied to the lungs, Dickson and co-authors suggested in a 2014 paper.[18]

Now the distance that matters is how deep in the lungs you go, hence how far from the source of microbes. Other factors include the size of the bacterial population in mouth and nose, and how well cilia and immune cells are working. They suggest that a model like this can help assess species richness in the lungs, but not the total microbial population. That depends on the rate of reproduction once a bacterium has found a spot to colonise.

It also helps to throw away the idea that the respiratory tract is only compartmentalised into 'upper' and 'lower' bits. There are really many sub-regions, each with different local conditions that can affect what may live there. The internal surface area of the lungs – around 30 times that of your skin – varies along a whole series of gradients, including temperature, oxygen levels, acidity, and the cellular structure of the lung epithelium. All of these matter if you are a roaming bacterium. The precise effects remain to be explored, but 'a consistent relationship between an environmental gradient and a type of microbe would be a powerful argument that bacteria are not only present in the lower airways but are also actively reproducing and susceptible to selective pressure'.

The ecological view alters the understanding of disease as well as what is defined as normal. If the healthy lungs have detectable microbial populations then it is too simple to view lung diseases – most notably the mass killer pneumonia – simply as ill-effects of interlopers.

Instead, says Dickson, pneumonia is the outcome of a disruption in a complex ecosystem, one that can often adapt to a species that has the potential to become infective. The picture that emerges from microbiome studies is that the species identified as the 'cause' of any particular case of pneumonia is only one of many that are present, each facing a combination of promoters and inhibitors of growth. Species that turn pathogenic and cause acute infections in some people are often found existing quietly in the lung microbiota of others with no infectious symptoms.

He concludes that 'the lung microbial ecosystem is a complex adaptive system within which pneumonia is an emergent and disruptive phenomenon.' That sounds grand, but does it actually affect the bottom line – how this globally important condition is diagnosed, treated or even prevented? He thinks it does. It directs attention to different features of the course of pneumonia.

In the old picture, pneumonia takes hold when a big dose of an infectious bacterium enters sterile lung tissue and overwhelms the normal defences. But if there are always a collection of species in the lungs, interacting with the host cells and with each other, then the complex system that results may be susceptible to a big change, like sudden growth of an infection, triggered by quite small shifts in the checks and balances that maintained the population mix as it was before.

This is consistent with one of the worst features of bacterial pneumonia, that it comes on abruptly, in hours or days. How could that happen? Dickson suggests there are many possible feedback loops that might produce a sudden population shift to very large numbers of a single species. Suppose, for instance, that excess growth of a harmful species is kept down by both a limit on nutrients and a low-level inflammatory response. If the inflammation goes up a level it can cause cell damage that causes nutrient-laden fluid to leak out of nearby tissues. This promotes bacterial growth, that causes more inflammation, that increases the nutrient supply still further. The pathogen population explodes like an algal bloom in the ocean. This is a mechanism we will meet again in other parts of the body.

That is just one idea, but it illustrates how to think about the disease now we know there is a normal lung microbiome. There are other possibilities. Production of mucus, for instance, is ramped up if there is infection, but it can provide a food supply for bacteria as well as a clearing action.

Recognising a new lung ecology also makes sense of some well-established clinical observations. People being kept alive on ventilators in hospital are at risk of pneumonia, and the risk increases if they have been treated with antibiotics. Hard to make sense of *that* if the lungs are normally sterile.

And the importance of normal lung microbiota may be reinforced by early trials of probiotics – bacteria thought to be healthy – for patients on mechanical ventilators. A randomised trial of 146 patients on ventilation saw pneumonia halved in the probiotic group.[19] In this case, the microbes they received were swallowed (actually introducing probiotics into the lungs is still a step too far), but perhaps some found their way into the respiratory system, for the same reason that there is a lung microbiome in the first place: whatever goes down your throat always has the lungs as a possible destination.

Even your eyes ...

Your eyes have a built-in bacterial cleansing system, combining washing and wiping, and it was long assumed their bacterial load was light unless there was some pathogen breeding where it should not.

Detailed DNA analysis has overturned this assumption, too. University of Miami ophthalmologist Valery Shestopalov found that the mucosal surfaces of the eye – the cornea and the inner surface of the eyelids – were no exception to the general rule that bacteria like body surfaces, internal or external, that are warm and moist. He quickly established that there were as many as 300 kinds of bacteria on the conjunctival surface (inside the eyelids), four times as many as had ever been cultured and quite a few of them entirely unknown.[20]

Cue the Ocular Microbiome Project, led by Shestopolov, which reported preliminary results in early 2014.[21] The aim was to get a closer look at the normal population living on healthy corneas and eyelids, and to consider the effect of an unprecedented modern development – contact lenses that float on top of the corneal fluid. The ocular microbiome team want to find out whether the normal corneal flora maintain an equilibrium which helps prevent infection by less desirable colonists, and whether contact lenses disturb this population.

The full details await publication, but it seems there are a dozen bacterial genera that predominate on the normal conjunctiva, and the same kind of diversity, though not all the same organisms, on the cornea itself. As elsewhere, normal healthy corneas can have widely differing bacterial populations, which will make sorting out the differences between health and disease more complicated.

However, infections of the cornea do go along with a reduction in bacterial diversity, which can be picked up before the infection itself is obvious. And the project is also logging the bacteria found on contact lenses, which can become coated with a biofilm that supports a community of microbes that is different from, but overlaps with, those found on the cornea normally.

We already know that contact lenses can lead to infections that cause irritation, or worse, in a significant minority of wearers. Whether this relates to their effect on the eye's regular bacterial colonisers, and if so how, remains to be seen. For now, I am glad that an intuitive disinclination to put bits of glass or plastic in my eyes has always deterred me from substituting contact lenses for the glasses I normally wear. I daresay my glasses are teeming with bacteria, too, but they are not competing for living space with those in my eyes, as far as I know.

Big and little

None of these bodily ecosystems is insignificant, and each of them will offer lifetimes of study for some researchers. There is more to say about each of them when we look in more detail at health and disease. Our whole microbiome has a long cast list, but a minor player can still be given a line that affects your interpretation of the whole play.

Still, there is one star, one vast reservoir of microbial life that we have been skirting round. The gut relies more heavily on microbes to do its job than any other organ or tissue. It is where they lie nearer the heart of our physiology, and cause more problems when things go wrong. That is partly because they are far more numerous there than anywhere else. If our other microbial collections are large and full of interactions that have to be teased out slowly and carefully, the gut microbiome is really dauntingly complex. After this round-up of the smaller communities in our microbiome, it is time to tackle the big one.

5 | **The big one**

Open a regular-size can of soup – no need to take Howard Hughes-style precautions – and dump the contents in a bowl. Do it again, once or twice. The serving of organic broth in front of you is roughly the same as the volume of bacteria in your colon.

This is not a new fact, but it keeps coming back to me as I try to make sense of the microbiome. Microbes are small and will happily live small. A decent colony of bacteria can maintain itself as a spot, a smear, a smudge or a thin film of life. It is easy to forget that fast growth means they can quickly generate a large mass, given the opportunity. Industrial fermenters brewing up a million litres of microorganisms are not uncommon.

Modern microbiome studies have sharpened my appreciation that my body furnishes a collection of ecological niches for other organisms. But most of them remain comfortingly microscopic. The gut is a different proposition. If you could remove all the microbes from the rest of the body and combine them in one sample, they would leave room in a teaspoon. The microbes in my digestive tract would need a ladle.

Sheer mass means you can reasonably think of the gut microbiome as another organ. If so, it is an important one – as active metabolically as the liver. And an odd one to contemplate. This organ is made of cells descended from ones that originally came from elsewhere. A large proportion of them are expelled each day. Can I assume, as I do for all my other organs, that it is looking after my interests? And what, actually, is it doing?

While the other sites and species that are part of the total human microbiome have their role to play, it is the gut microbiome, and especially that of the colon, that seems the most important to try to understand, and that has the most far-reaching effects. The next three chapters look at what we have learnt so far.

Down the hatch

There is one obvious way into the gut: via the mouth. But a bacterium that aspires to join the gut microbiome has a long way to go and plenty of possible destinations.

The human intestinal tract is continuous, but has quite a few distinct regions. The three main ones are the stomach, the small intestine, coiled in the abdomen, and the large intestine or colon. If you consider it simply as an extended tube, the whole thing would stretch out to seven metres in an adult.

The number of bacteria increases massively as you follow the path that a mouthful of food takes. After the microbially rich region of the mouth, the stomach contents – a mixture of mushed food, saliva, and highly acidic secretions that help to break down incoming protein – only support about ten microbial cells in every gram. By the time you reach the duodenum, the first portion of the long small intestine, the population has gone up to 1,000 per gram, and there's a further 10,000-fold increase along its length, with the final part of the small intestine yielding 10 million microbial cells per gram. The largest leap, however, is between the small intestine and the colon. This last major stretch of our gut, once thought of as a rather dull pipe in which fluid was reabsorbed from waste matter on the way to becoming faeces, nourishes a teeming mass of a million million (10^{12}) microbes per gram.

All this other life occupies a space whose evolved organisation is itself a microscopic wonder. The intestines, large and small, have two jobs, with conflicting requirements. Unlike the microbial playgrounds of the skin, the intestinal surface cannot just act as a barrier. All the small molecules produced by digesting food, the main local enterprise, must be absorbed into the bloodstream so they can be used where they are needed. That means the gut wall needs to be thin and have as big an area as possible. It is easy to see how thin: the surface layer, or epithelium, is around ten microns across, about ten times the size of a typical bacterium. It is harder to estimate the area because the surface is convoluted on several different scales. If the intestinal wall, the thin layer, were flat it would cover less than one square metre. It isn't. Innumerable small projections, or villi, stick out into the internal space, the lumen. Each of those villi maintains the single-cell layer of epithelium. But cells have outer layers, too, just visible under light microscopy as a layer that looks vaguely bristly, called the brush border. Use an electron microscope and the border is seen clearly as a further set of microvilli, which adorn the lumenal side of the cell membranes.

Figuring the area of all this is like trying to come up with a surface area for your fluffiest bath towel, but the figure that anatomists agree on is between 200 and 250 square metres, roughly the size of a tennis court. That may not be totally accurate, but it is in the right, well, ballpark. Take home message: it is really big.

That is good. The gut, like the lungs with their finely branched alveoli for gas exchange, needs a big expanse to ensure efficient molecular transfer. However, here we hit the other conflicting requirement. The gut is full of bacteria as well as food, and we want to keep them from crossing into the bloodstream. We tend to think of our skin as the main barrier to entry deeper into our tissues, and it does have an important barrier function. But the gut is far larger, and confronts far more bacteria, more of the time. How does it do it?

There is a barrier, or course. The epithelium, like all biologically active boundary layers, manages the molecular traffic while barring passage to larger items like microbial cells. And neighbouring epithelial cells are sealed together by a protein mesh in the closest arrangement multicellular organisms use, known as a tight junction. That helps too.

The full answer, though, involves the immune system. As the gut is at the heart of metabolism, and dominates the microbiome, it is also the largest operating space for all the molecular and cellular bits and pieces that maintain immunity. The presence of trillions of bacteria inside the gut is probably the main influence on how immunity develops, both in evolution and in each new individual, an influence we are only just coming to terms with. The way that is transforming our understanding is so important it needs a chapter to itself (see Chapter 7).

Meanwhile, let's look at the microbial contents of the most densely colonised region, the colon. It is the most complex microbial ecosystem in the body, and in terms of cellular density and diversity probably the most complex ecosystem anywhere. It dominates all the surveys of numbers of different microbes and genes. The more people who are studied, the more diversity is revealed. The first reported gene microbial catalogue, compiled from 124 people, recorded 3.3 million genes in the gut. The latest,[1] combining results from nearly 1,300 people from America, Europe and Asia, has pushed the total to over 10 million.

So, yes, this continuous culture system, in which nutrients are fed in and microbes leave at a more or less constant rate – a few trillion living and dead bacteria are flushed away every time you empty your bowels, so the mass estimates are very much an average – is fearsomely complicated. And as I said in the previous chapter, there does not seem to be a core human microbiome. Nor, as the gut flora dominate the total so heavily, is there a core human gut microbiome.

There may be some useful simplifications, though, when we try to figure out what is important down here in the colon. One is controversial, the other less so.

The one in dispute is the idea that the gut microbiome comes in a small range of broad types. It arose from the first analysis of the mainly European metaHIT (for metagenomics of the Human Intestinal Tract) project, which ran at the same time as the US Human Microbiome Project. In 2011 they reported they had found three groups of people who could be distinguished by the microbes in their gut. The idea was that each distinct community of microbes amounted to a stable ecosystem. Their microbial signatures did not seem to be related to any other common variables like age or sex, and were easy to read. Each had a big portion of one kind of microbe – either *Bacteroides*, *Prevotella* or *Ruminococcus*.

Cue much speculation about how this would lead to crucial diagnostic tests, or even to an aid to identification, like blood groups. However, this result was based on just 39 people, and the more samples that were analysed, the fuzzier the picture got – an oft-repeated story. The basic idea probably has something in it, though more in terms of clusters than clear-cut types. A group of 35 chimpanzees in Gombe National Park in Kenya had enterotypes, too, similar in some respects to ours.[2] The same chimp sometimes had a different enterotype when sampled a year later, though, so their long-term stability is in doubt. The changes did not follow any obvious pattern. One family of three chimps – a male and female and their mother – all had different enterotypes on first sampling. Each of them changed later in the study, but in such a way that all three still remained different from each other.

I'll say more about evolution of the microbiome in Chapter 6 but it does look as if a tendency to maintain enterotypes, or at least clusters of species, goes back a long way, at least to the common ancestor of humans and chimps. And house mice have it too, according to a report in 2014 – though only two enterotypes showed up. It will probably resolve as a clustering along particular gradients. But if that proves the best way to describe what is happening, the clustering in human guts differs from those found so far in other animals. Neither mice nor chimps have so far shown a *Prevotella*-dominant enterotype.

Much of the subsequent work on gut microbiome population and what people – or mice – eat has come in an effort to figure out how microbes affect obesity, which I take up in Chapter 8. However, deeper analysis of the Human Microbiome Project data has gone on, and it does produce other evidence suggesting that microbiomes at all of our body sites probably exist in a few common states, and that their typical clusters of microbial types, whether in the gut, the mouth or the vagina, can all be found in healthy subjects. An intriguing paper in this vein from a team at the University of Michigan catalogued all these microbiome types from HMP samples,[3] and reaffirmed that, as co-author Patrick Schloss put it, 'there is no one healthy human microbiome'. There were some slightly mystifying correlations, like those between the mouth, vagina, and elbow creases or behind the ears. However, one that stood out was between the gut (or the stool samples) and communities in the mouth. In fact, claims Schloss, 'The type of bacteria you have in your mouth can predict the type you have in your gut'. Maybe that is good news for those who hold that you can get healthy by eating 'good' bacteria.

The other, less controversial simplification pertaining to the gut microbiome comes through looking at its dizzying diversity. There are two aspects to this: the number of species any one person carries, and the differences between people.

Diversity in any one colon is large, typically hundreds of species, but it is often the case that there are a few species – perhaps only a couple of dozen – present in large numbers and a lot that manage to maintain a much smaller population. We do not know whether we can ignore some or all of this long tail of the distribution. It may, for example, provide a resource, a pool of genes to dip into when need arises. The rapidity of bacterial growth when conditions are favourable means that a minority group can rise to dominance fast. But this situation does invite attention to the common majority species, which helps identify priorities for follow-up research.

Another simplifying finding emerged from the first HMP survey, and has stood up since. There is lots of variation in different people's microbiomes when you map the species that turn up in their stool. But move beyond simple 16S rRNA typing and analyse the whole pool of DNA a different way, paying no heed to species, and the picture gets much neater.

To see that, you line up all the bits of DNA that look as if they are functional genes, identified from their characteristic control sequences. Ignore the ones that have no obvious match in the databases. Finding what they do must await later research. But for the large number that either have known functions, or look as if they do, group them into functional units. Most genes code for enzymes, and metabolic processing usually involves chains of chemical reactions in a sequence, each encouraged by interaction with a different enzyme molecule. These enzymic co-conspirators tend to go together in sets of genes that have been dubbed 'metabolic modules'.

Map the occurrence of these modules and there is much less variation in microbial gene function between samples than their species differences might suggest. Each gut microbiome, it seems, may not have the same bacteria, or the same genes, but it has genes that do roughly similar things. The same is true at other HMP sampling sites, but the gut results show the most stable genetic mapping, suggesting that the digestive functions of gut bacteria are carefully stewarded.[4]

Analysis of the accumulating data goes on, and new compilations fill out this picture without altering it radically. The expanded catalogue of 10 million bacterial genes found in human gut samples provides a new foundation for this work. One superbly clever use of it is that once all the information has been compiled, even though the database lists only individual genes, you can use them to dive back into the total pool of samples and identify hitherto unknown bacteria. Genes never occur in nature as isolated scraps of DNA, but flock together in chromosomes – bacteria have just one of them each. If computer matching shows that a collection of genes are always found together, that means they are joined together in one or other bacterial species. This 'co-abundance' approach does indeed lead to lists of genes that match the genomes of species that are already in separate databases of bacteria that have been sequenced whole. That validates the algorithm the researchers are using. Then it rewards them by showing up previously unseen gene combinations that represent completely new species, whose role in the community can then be investigated. It is yet another way of sifting information to make the invisible come into view.

The overall catalogue allowed a more refined analysis of individual diversity that goes some way to redeeming the notion of a core human gut microbiome – in terms of functions, if not species. The widescreen analysis of the complete set of genes that have ever been found in a human gut, anywhere, which came up with that impressive figure of 10 million, also allowed the researchers to show that around 300,000 were present in almost everyone who has been sampled. Each of us only carries 600,000 or so bacterial genes from that pool of 10 million at any one time. So perhaps half of anyone's healthy gut metagenome is shared with everyone else.

This all makes intuitive sense. Even if bacteria are forming an ecosystem by random colonisation and competition to survive, the potential nutrients in one colon, for instance, will be fairly similar. If one species does not come along that can exploit some of the undigested goodies, then another probably will, using the same or similar enzymes. Once either is established, it will be harder for another that wants to make its living in the same way to get established. Result: diverse ecosystems but similar collective metabolism. Which leads directly to the general answers to the question that immediately arises when contemplating all the trillions of bacteria down there in the gut. What are they all doing?

Services rendered

If you manage an organisation, there are important decisions to make about what capabilities to maintain in-house, and when outsourcing gets the job done more efficiently.* So also for organisms. And it turns out that our rather large organism has outsourced a lot of jobs to much smaller ones.

Gut bacteria are often referred to as commensal organisms. As I said in the Introduction, it is an ecological term (it means 'eating at the same table') for organisms that live on some other creature but do no harm.

There may well be commensal microbes in the microbiome, but gut bacteria definitely aren't simple commensals. The whole set-up revolves instead around mutualism, in which both parties benefit. We provide food, a nice place to live which is kept at a comfortable temperature and, when we expel some of them in faeces, the dispersal all organisms need to ensure their long-term prospects. In return, bacteria carry out a host of important tasks that our own genome doesn't need to provide for.

They include simple digestion, working on the as yet unused parts of food to release energy, some for themselves, some for us. But they also make a wide range of small molecules, including vitamins, on which we rely. And they help dispose of a range of toxins, and – in modern times – to metabolise many drugs.

The most prominent role, so much so that the colon has

---

* Or so I hear. I'm a writer, so organisations aren't my thing.

been dubbed a second stomach, is digesting complex carbohydrates. Chewy plants build their cells using many large molecules that pass through our stomach and small intestine unscathed. Some of the starch we eat, the portion known somewhat non-committally as resistant starch, also eludes breakdown in the stomach or small intestine. When this partially digested or undigested material arrives in the colon, bacteria get to work and finish the job. Their enzymes break down this plant material into small molecules that can be used by our own cells to produce energy. Between 10 and 15 per cent of an average adult's food energy is obtained this way.

This bacterial assistance comes from a large battery of enzymes, dealing with lots of different foods. They include all the familiar types of dietary fibre, from oat and wheat bran to the complex carbohydrate inulin found in onions, garlic and asparagus. Plants build their cell walls from molecules that are designed to last, made by linking soluble sugars into complex branched chains that are anything but soluble. Pretty well all fruit, vegetables and grains contribute substances to the diet that reach the colon largely unchanged.

Many of the complex starches and other carbohydrates involved are worked on co-operatively by different species of bacteria. And some of the microbes that have been evolving to work with us are formidably well equipped to deal with whatever comes along.

The champion from those that have been studied in detail so far is *Bacteroides thetaiotaomicron*, a species found only in the gut. This bacterium alone has genes for 260 different carbohydrate-degrading enzymes. Our own human genome musters a mere 95 such enzymes even though it has about 1,000 times more DNA. We find it more convenient to let a super-versatile bug make the rest for us. The bacterium switches adroitly between enzyme combinations according to what food is available, responding to a combination of what its host is eating, what enzymes host cells make, and what metabolites other bacteria present may use. The diverse enzyme genes are supplemented by another 200 bacterial genes thought to code for proteins involved in binding or transporting starch. This is a bacterium seriously dedicated to digesting the indigestible. That makes it a pillar of the community in the gut – providing extra value from foodstuffs for us, and helping other bacteria that make use of some of its fermentation products to sustain their own populations.

It has also evolved with us, and other mammals, to ensure a mutually profitable co-existence. Experiments with germ-free mice exposed to B. *thetaiotaomicron* show that when it is present epithelial cells in the mouse's intestine increase production of complex carbohydrate with a particular sugar at the end of a chain, which the bacterium can chop off and use for food. Conversely, epithelial cells' production of a molecule so well-fitted to the bacterium's tastes encourages it to colonise the intestine.

The relationship is still tighter. Adult germ-free mice have fewer than normal capillaries in the tissue underneath the intestinal surface layer. Introducing the bacterium starts the blood vessels growing again, helping their new host absorb the nutrients it will supply by applying its battery of enzymes to otherwise indigestible carbohydrates.[5]

Recall that bacteria are good at swapping genes. This is one way one species can acquire so many enzymes for a specialised diet. We can get an idea of how B. *thetaiotaomicron* has accumulated some of these enzymes from a study widely reported in

2010 showing that a related gut *Bacteroides* species has picked up an enzyme called porphorynase from a marine bacterium in the same genus. The enzyme is common in the gut microbiome of the Japanese population, many of whom are ingesting the complex polysaccharide porphyran with the seaweed in their sushi. The evidence may be circumstantial but the obvious conclusion is that in Japan, but not in North America, the gene for porphorynase gets picked up by the gut species from marine bacteria that may be eaten along with the seaweed.

It is easy to imagine this scenario playing out with other plants. Any food that needs fermenting is likely to come, at least some of the time, with a dose of bacteria that have responded to the challenge of breaking down the complex molecules the plant has taken such pains to synthesise. Bacterial genetic exchange does the rest. There is good evidence that there is more of it when bacteria are in the gut – where genetic exchange may be 25 times as common as it is when similar microbes live elsewhere.

The metabolic virtuosity of gut microbes extends in other directions, too. They break down polyphenols in cocoa powder to make small molecules that may have an anti-inflammatory effect on blood vessels, producing the welcome possibility that eating dark chocolate might be good for you. More generally, they do not just break things down. Microbial enzymes help produce a large variety of essential small molecules, including B and K vitamins, some neurotransmitters, and basic feedstock for all cellular enterprise, like amino acids. Some of these are also made by our own cells, but many are not. A rough estimate frequently quoted is that as many as a third of small molecules being ferried round the body in the bloodstream originate with gut bacteria.[6]

As with the many kinds of bacteria, there is little to be learnt from simply trying to enumerate all the molecules they help us make. But if we want to take stock of what it might mean to understand all our interactions with gut microbes, it is worth stepping back for a while from the challenge of taking in their total contribution to consider, what can just one molecule do? The answer: more than you might think.

Molecular promiscuity

Small molecule seeks partner, impressive evolutionary history, can function as an energy source if required, GSOH.

Molecules do not post ads in the personal columns, but some still enjoy dalliances – brief but influential couplings – with a large catalogue of significant others. Evolution is a pilferer and if some small, fairly stable molecule has been around for a long time it keeps on finding new uses for it.

This is one reason that bacteria in the gut are much more than aids to digestion. They produce lots of small molecules, which are read as signals in many of our cells, tissues and organs. The networks that result seem endlessly complex. A full mapping of what interacts with what, if it were achievable, would end up looking like a wiring diagram for the internet. If I had one, I would immediately be hoping for news of some general principles and higher-level properties.

Still, one crucial level of interaction here is molecular. So let me try to get an idea of how a collection of tiny organisms in my gut can affect a much larger human by investigating a single molecule.

Meet butyric acid. In the schematic formula chemists use (letters are atoms of the designated element – carbon, hydrogen, oxygen in this case – lines are bonds linking them together), it looks like this:

$$H-\underset{\underset{H}{|}}{\overset{\overset{H}{|}}{C}}-\underset{\underset{H}{|}}{\overset{\overset{H}{|}}{C}}-\underset{\underset{H}{|}}{\overset{\overset{H}{|}}{C}}-\overset{\overset{O}{\diagup\diagdown}}{\underset{O-H}{C}}$$

It is one of a class of compounds called short-chain fatty acids. The acidic bit, that COOH on the end, stays the same throughout this class. It uses up three of the four bonds that keep a carbon atom happy, those in turn being the key to carbon's ability to make chains and generate almost endless compounds that make it the key element for life. Use the fourth to grab a hydrogen atom, H-COOH, and you have formic acid, the irritant in insect stings. More benignly, a 'chain' of two carbons gives you acetic acid, in the vinegar bottle in your kitchen. As you can see, butyric acid strings four carbon atoms together. These chains can be quite long – hexacosanoic acid has 26 carbon atoms in a chain – and can have lots of other features I do not need to bother with. The short ones are not very fatty, and butyric, like the others with just a few carbons, is soluble in water. It is an acid because the hydrogen on that OH group can break free as that chemically most simple entity, a positively charged hydrogen ion – otherwise known as a proton. That leaves a now ionised oxygen atom on the larger molecule with a negative charge. This negative ion is known as butyrate.

And butyrate is being churned out by many of the bacteria in my gut. At first look, this is for the reason I've already given. Producing short-chain fatty acids allows me to make much better use of what I eat. The evidence from germ-free mice does seem to confirm that the bacteria do that. Such deprived rodents typically need to eat 10 per cent more than mice with a normal microbiome to maintain the same body weight. That is an important clue to our new view of the dietary fibre we are always being urged to swallow. Complex carbohydrates, the main component of dietary fibre, tend to end up undigested in the colon. We used to think they were good for you because they somehow helped the colon work more smoothly, adding bulk to its contents. Eat fibre-free meals and you would risk constipation and, eventually, colon cancer.

Fibre's fate turns out to be far more interesting than as a bulking agent for turds.* If the right bacteria are there, the big molecules are broken down by bacterial enzymes, producing short-chain fatty acids. They, in turn, can be used by our own cells to produce energy. Acetate, which is typically produced at about three times the rate of butyrate, passes into the bloodstream, and is used in muscles and liver in a similar way to glucose. Some butyrate is also absorbed from the colon and used in the liver. But its metabolic importance begins right in the colon, with the rapidly dividing epithelial cells there hungry for butyrate and liable to start digesting their own contents if they do not get enough of it.

That would be a nice, neat scheme, with bacteria providing a convenient energy source for needy human cells nearby. But a molecule of butyrate that escapes being used for food by body cells can do all manner of other things. There seem to be receptors that recognise it, from its shape and the way electrical charge is distributed between its atoms, all over the place. How many? We probably do not know them all yet, but let us visit

---

* Though it does that, too. I mention this because it prompts mention of the Bristol Stool Chart – a seven-fold classification of the varied textures of human faeces. As a resident of the great city of Bristol, I feel this contribution to knowledge should be more widely appreciated. It is used as a proxy for digestive transit time.

a few. Molecular interaction in living systems is often fast and
fleeting. Imagine a single molecule, in a fluid space crowded
with others, being constantly jostled and bounced around, and
with its own atoms vibrating or even rotating around their
inter-atomic bonds. It may well clasp hands briefly with some
receptor or recognition site, then get nudged onward, back into
the flow. If a newly arrived butyrate ion in the colon was handed
a list of 'twenty things to do before you get metabolised', it
might begin something like this.

Find and bind to a G protein-coupled receptor. These are
an extended family of cell-surface receptors that do what their
name suggests. They span the cell membrane and bind some
small molecule outside the cell. That small change provokes
the receptor into changing its shape, and it then activates one
of a class of proteins – G proteins – that relay a signal inside
the cell, producing a wide range of effects.

Lots of signal systems in cells work like this, and there
are hundreds and thousands of different G protein-coupled
receptors in our tissues, as well as receptors in the same family that work through different relay agents, so it is no surprise that some of them bind butyrate (and, for that matter,
acetate). Their profusion means they get simple numerical
labels. In this case, the first one our molecule encounters is the
one labelled Gpr43, which is shaped to bind all three of the
common short-chain fatty acids, and helps keep inflammatory
responses damped down.

That done, our molecule of butyrate moves off that receptor, and on to another, Gpr109a, which ignores the other
short-chain fatty acids and only grabs hold of butyrate
(although, as cell biology is full of criss-crossing pathways,
it can also respond to vitamin B3 – niacin – another product

of some gut bacteria). This receptor has a similar anti-inflammatory role in the colon when activated. It seems to reduce the chances of colon cancer as well. Incidentally, in a typical interconnection – one of those loops-inside-loops of feedback that help complex communities of cells self-organise – the production of the receptor in the colon is also boosted by the presence of gut bacteria. Is this another effect of butyrate? We do not know yet.

But this, too, is a brief encounter, and our high-performing molecule of butyrate then wafts away, to bind instead to receptor Gpr41, which signals cells to increase production of leptin, a crucial hormone in control of appetite, and fat metabolism and storage. Finally, it is picked up more assertively by yet another type of receptor. This is a transport protein that carries butyrate inside a body cell, in this case a colonic epithelial cell. Once inside, it is released and can interact with new partners. One well-studied effect of butyrate inside cells is inhibition of an enzyme that removes acetyl groups from proteins called histones that package our DNA. This gets your attention when you learn the additional fact that raised activity of this enzyme is one characteristic of cells from colon cancers.

That could be one endpoint of this molecular excursion, but there are plenty of others possible. If the cell ferrying the butyrate inside happened to be a T-lymphocyte, it might be induced to differentiate into a more specialised kind of immune cell. There are also transporters that shunt butyrate across the gut epithelium altogether so that it enters the bloodstream. It can then go almost anywhere. Similar transporters, some suggest, can carry short-chain fatty acids into the brain and nerve cells. There might even be a connection between that detail and the fact that experiments in mice suggest large doses of butyrate can have an anti-depressive effect, a finding we'll come back to in Chapter 9.

But let's end this particular invented journey here. It summarises only some of what we know about butyrate and what it can do. What we know so far is certainly not the complete story. But it can stand for many other, similar stories of molecules that have been cast in a whole series of roles as the coordination of our body's systems, and those of all our evolutionary predecessors, has developed.[7]

There are scientists trying to unpick all the other stories in this kind of detail, and any of them offers a reminder of the finely poised interaction and coordination that goes with trying to operate an organism with a trillion cells. What is different in the superorganism is that the whole system now includes trillions more cells that operate, in some degree, independently, and have interests that do not coincide precisely with those of our other body cells. Their activity nevertheless has to be orchestrated along with that of all the other cells if the whole show is to go on. This leads to a continual molecular chatter, of which this sketch covers a small part.

I take two other morals from this little story. No small molecule is ever likely to do only one thing. In fact it is quite likely to be involved with different systems in ways that can, at times, have seemingly contradictory effects. And not only is one molecule implicated in fine-tuning of many body systems at once; the signalling networks its action modulates will be entangled with many others that would have to be elaborated equally carefully to have any clear idea of the likely final results. All the effects of my hypothetical, hyperactive single molecule of butyrate depend on the precise cellular circumstances. These are things to bear in mind when trying to understand whether new research findings about the microbiome and its effects mean what someone claims they mean.

We need to get back to the higher levels of the microbiome, of ecosystems and interactions. But while we're down here, being devoted reductionists and trying to examine things one at a time, let's make time for one more story with a single protagonist. This time, it is a single bacterium.

Good germ, bad germ

One lasting impression from the welter of new results from DNA analysis is that there are an overwhelming number of kinds of bacteria that can appear in the human microbial population. That variety, and the even larger variety of the 10 million-odd genes all those bugs have at their collective disposal, is one of the big things to try to understand.

Nevertheless, there are things to learn from homing in on just one bacterium. As with the endlessly fruitful lab studies of *E. coli*, learning as much as possible about a single bacterium can yield general lessons about how they interact with us.

Take the intriguing story of *Helicobacter pylori*. It is particularly suitable for single-species studies as it is acid tolerant and can live in the stomach, where the bacterial population is much lower than in the rest of the gut. That may not make it the best guide to the way we interact with the rest of the microbiome, but it does illustrate some important things about our relations with bacterial species.

The story has had some remarkable twists and turns, involving overturning the scientific and medical consensus not once but twice in the space of 30 years – and the second time required demolition of an opinion firmly cemented in place by the first.

The most celebrated tale concerning *H. pylori* is the work that established that it causes ulcers. Right up until the 1970s, everyone thought these painful and sometimes dangerous erosions of the stomach wall were due to too much acid, an effect brought on by stress. Then an Australian pathologist, Robin Warren, noticed that there were bacteria in the stomach, colonising the mucus layer lining the stomach walls, and that they went along with inflammation. It turned out that the bacteria, which had been observed in the 19th century and then so thoroughly forgotten that doctors were taught that the stomach was sterile, were strongly associated with ulcers.

The association was strong enough to make the bug a reasonable candidate for a pathogen that obeyed Koch's rules. In 1982, it was cultured in the lab, a tricky feat that had eluded earlier microbiologists. Then it was shown: that you can recover *H. pylori* from people with ulcers, and from the larger number with gastritis, or inflamed stomachs; that swallowing the bacteria brought on an attack of gastritis – an experiment that made Warren's collaborator Barry Marshall famous; and, most impressively, that treating *H. pylori* infection with drugs helped clear up ulcers. Result, a Nobel Prize in 2005 for Warren and Marshall and a legion of doctors convinced that this new (to them) bacterium was a dangerous pathogen. If the stomach turned out not to be sterile, then it certainly ought to be. The only good *H. pylori* was a dead one. Bring on the antibiotics.

However, while the extreme environment of the stomach keeps the microbial population small and simple, this turned out not to be nearly as simple a story as Koch would have wished. Martin Blaser, of New York University, noted early on that not everyone who carried *H. pylori* got ulcers. In the late 1980s he began to study the organism more closely, and his group soon found that there were several variants, and that people infected with them carried antibodies in their blood. A blood test soon followed.[8]

He went on to dissect the varied types of *H. pylori*, which differ in the way they interact with epithelial cells in the stomach, and show why some were more likely to bring on ulcers than others. He and others also showed that the bacterium increased the risk of stomach cancer, a major cause of death. More incentive to prescribe antibiotics – we must eliminate this deadly bug whenever possible!

But that was not the complication. What Blaser realised was that these links with disease only showed up because not everyone carried *H. pylori*. It looked, though, as if this was a recent development. The 19th-century reports, and contemporary studies in Africa and Asia, showed that essentially everyone had the bacterium in their stomach lining, or had antibodies in their blood suggesting that it must be there.

More thorough recent investigation suggests that it is a very ancient companion. Every mammal appears to have a related species in its stomach, evolved along with its host, and we can show that people have carried it for at least the last 100,000 years, probably longer. *H. pylori* lives only in humans, and must be passed on after birth, but in less hygienic times most children picked it up by the time they were ten.

Then came modern sanitation and, later, frequent antibiotics for childhood infections. The incidence of *H. pylori* infection began to fall steadily. It does not pass easily between adults, and children may get it only if their mothers or siblings carry it. The result is that the microbe gains a foothold in a smaller proportion of people in each generation. Blaser estimates that the vast majority of people born in the USA in the early part of the last century lived with *H. pylori* in their bellies. For people born after 1995, the incidence has gone down below 6 per cent.

And a good thing, too. Those children and young adults will be at much lower risk of painful ulcers and often fatal stomach cancer.

But not an entirely good thing. Ulcers happen to some grown-ups, and stomach cancer does not usually appear until middle age or later. If all children throughout history until now harboured a bacterium that is specifically adapted to live in our stomachs, does it have any other effects, perhaps even good ones? The first indication that it might came from another investigation of Blaser's, looking at whether people who suffer from severe acid reflux, or heartburn, had more *H. pylori* than normal.

The result was the exact opposite of what the researchers expected. People who did *not* have *H. pylori* had twice the chance of getting the troublesome extreme of heartburn known as gastro-oesophageal reflux disease. This leads to painful daily episodes in which part of the stomach contents pass back up towards the throat. If that goes on it can cause scarring and worse. The ultimate result is adenocarcinoma which, like acid reflux, is on the increase.

Blaser's and other groups developed this work, finding that eliminating *H. pylori* with antibiotics often led to acid reflux. And there was what looked like a piece of pure biological perversity in their later results. *H. pylori* strains that make a protein called cagA, which can damage epithelial cells, are more likely to lead to ulcers and stomach cancer, but also offer a stronger risk reduction for reflux and adenocarcinoma. The reason for this odd swings-and-roundabouts pathological profile may lie in the way production of stomach acid is regulated, although the mechanisms involved are not really understood yet.

And there is more. It turns out that children who still pick up *H. pylori*, especially the ulcer-associated strains, are less likely to develop asthma, another condition that doctors are seeing more and more often. The same goes for hay fever and, it turns out, for a range of other allergies.

Understanding why that might be involves a look at the costs and benefits of inflammation, and the way our resident bacteria interact with our immune systems. This is discussed in Chapter 7.

But the multiple effects of *H. pylori* have generated a large research effort aimed at tracking its associations, whether positive or negative, with all sorts of other conditions. A 2014 review of current work references nearly 140 papers, and discusses things as varied as pancreatic cancer, anaemia, liver disease and arterial blockage.[9]

The associations here are mostly weak or ambiguous, and the research goes on. But the *H. pylori* saga demonstrates that, for at least some bacterial species, there is no clear division between a harmless or even a helpful organism and a pathogen. Sometimes it does one thing, sometimes the other. Which it is may depend on small differences between strains, genetic variation in the host, and lots of other personal and environmental factors that tend to yield a fuzzy picture of causes, if any at all. To make it clearer you would have to round up a really enormous study group, follow them for a long time, and analyse the hell out of the statistics you end up with. And that is just for one bacterium, albeit one that looks like a pretty important one.

My very own

Is there *H. pylori* in my stomach? I don't know. I do know that I used to be a touch asthmatic and I have never had an ulcer so it is tempting to assume not. But I do have a heck of a lot of microbes in my gut. The way to try to learn useful things about the microbiome is to move between the tiny details of action of a single molecule or a single bacterium and the zoom-out view that takes in a whole community. So to finish this chapter, I report a peek at my own colonic conglomeration.

If you want, you can easily get your own microbiome surveyed. It might even be a contribution to science, helping to fill out our picture of how much individual microbiota vary and what can make them change.

There are a few outfits offering to take your samples and tell you what is in them. American Gut is a crowd-funded project building a picture of US citizens' gut microbiota, one person at a time. I tried uBiome, another early entrant in the field and another US-based effort, which says it aims 'to equip all individuals with the tools they need, in order to empower them to learn about the unique balance of bacteria in their bodies'.

Their project was also crowd-funded at the beginning, raising $350,000, and their website suggests it is a contribution to 'citizen science'. Their idea is that people will use data from uBiome's sequencing machines, and comparisons with all the other analyses now coming out, to design their own studies.

For now, the project is pretty straightforward. You order a uBiome kit for $89, and a box arrives in the mail with swabs and vials. For the sample I sent back to them I went for the simplest option, a sample of stool recovered from toilet paper, swabbed, dipped in the stabilising fluid in the vial, shaken and sealed. I could have included samples from my nose, mouth, genitals and skin at extra cost but the gut, or at least its end product, seemed enough to start with. The others can wait.

The amount of poo involved seemed minute, and I

wondered how it could be enough when I packaged it up to send off to uBiome in San Francisco. Had I done it right? (It also feels odd sending a faecal sample through the mail, but apparently no one minds.) The rest of my input was online, registering the kit and answering a fairly simple questionnaire to establish a personal profile.

The project is still taking shape but, after some months' wait, at the time of writing I have access to the 'beta' version of the website displaying my personal data.

Before I dive in, I note the cautions on the site. The data is to help me learn more about my microbiome, but is not to be used for medical purposes. Also, 'some bacteria names may sound like infections or diseases'. This does not mean that I have (or don't have) an infection. As uBiome are using 16S rRNA sequencing, hence providing very broad-brush classification of bacteria, I knew that, but good to see the warning up there.

The data they do offer can still yield quite a lot of detail. The entry level is a nicely coloured diagram, presenting the following information about bacterial phyla in my gut – or at least in one tiny bit of my stool.

Most of the bacteria in my sample were *Firmicutes* – 74.5 per cent. This is appreciably above the average person in uBiome's dataset, who I am guessing are mainly North Americans, who has 61.6 per cent of this phylum. That average is based on samples from all the different body sites, but in this case is a near-perfect match for gut samples from healthy omnivores – which is how I would classify myself – at 61.67 per cent.

I was nearer the average for the less prevalent *Proteobacteria* (3.83 per cent compared with an average of 3.46), *Actinobacteria* (3.01 versus 2.72) but relatively low on *Bacteroidetes* (11.8 per cent versus 20.4, so only just over half the average). That disparity does not look significant when I drill down and look at the range of values for healthy omnivore gut samples, which goes from hardly any up to around 50 per cent, and all points in between. Just over 1 per cent of the types found were unclassified.

And then there is a surprise. There is a whole phylum I have never heard of, the *Tenericutes*, and 5.4 per cent of my gut bacteria fall under this heading. That compares with a uBiome sample average of 0.183 per cent.

Wow, an interesting discovery. My gut microbiome is unusual, in at least one way. Other gut microbiome samples are usually under half a per cent, (vegans average 0.65, but I am no vegan.) The range is pretty narrow, with very few outliers. I am *well* outside the normal range. Why?

And there the interest falters, for now. I have no idea what this high reading means. Neither, as far as I can discover, does anyone else.

I learn from uBiome that all my *Tenericutes* are from a class known as *Mollicutes*. These are simple bacteria that have no cell walls and are parasites that typically have a very basic genome, most likely arrived at by shedding as many non-essential genes as possible.

They are often found attached to cells in the lungs or genitals, and one of them is well known as it has perhaps the smallest genome of any cell, with fewer than 600,000 base pairs. *Mycoplasma genitalium* is an especially simple prokaryote that may come close to maintaining the minimal genome needed for independent existence, so is popular with genetic experimenters. But as its name indicates, it does not live in the gut.

I wonder whether this was a freak sample – it did come from a *very* small portion of one bowel movement. These tests have not been going long enough yet to give us much idea whether the analyses are repeatable – although science journalist Tina Saey reported very different results when she sent portions of the same sample of her own to uBiome and American Gut. In fact, she reported, the two gave more or less opposite results at the level of the proportion of *Firmicutes* and *Bacteroidetes* in faecal samples taken from the same section of the same piece of toilet paper. When she queried the companies, they explained how many things, from DNA extraction techniques to data-processing algorithms, could account for the discrepancies – a reminder that microbiome studies are a long way from standardisation.[10]

I am still wondering whether my own result would be confirmed if I paid up again and repeated the analysis today. Then I would want to know why this class of bacteria have taken a special liking to my gut. But one explanation could be that it just looks as though I am unusual because of a limitation in the current uBiome data set. When I turn to the wider scientific literature, I find that *Tenericutes*, though news to me, are not unknown in the gut microbiome. One comparative study in 2013 found that *Tenericutes* made up 12 per cent of the gut bacteria in the Bangladeshi children who gave samples, and 4 per cent in American kids. Suddenly, my result looks less interesting, just a bit on the high side for a Westerner.

All in all, I had fun looking at my results, and the full comparisons within the uBiome data set are a useful reminder of how varied personal microbiomes are. But the data do not tell me much more. This sort of personal analysis will need to get a lot more detailed before it is actually useful to the microbiome's owner (if that is the right word). There is more detail available now – my *Firmicutes* divide among *Clostridia*

(69 per cent of the total sample), and *Negativicutes*, *Lactobacilli* and some things called *Erysipelotrichia*, all at 1 or 2 per cent each. But all these classes still contain a further multitude of species, so investigation at this level still does not feel that close to getting acquainted with the kinds of bacteria that are actually living in my gut. On the other hand, the information that has come through from studies at that level suggests there is even more individual variation, perhaps partly because there are just so many bacterial species involved. It all makes the day when getting hold of your personal microbial profile, and having a clear guide to what it means, look anything but close.

That is one snapshot of one adult's gut microbiome, on one day. It is certainly well-populated, and impressively diverse, even if we only examine the bacteria and ignore other types of microbes. I also know, I think, that I started life without any microorganisms inside me. So, how did they all get there?

6 | A microbiome is born

Here we are, you and I, 21st-century adults, with our mature microbiomes. We can pore over ever more detailed snapshots of each of the ecosystems that form part of this assembly. But where *did* they all come from? There are two ways of approaching that question. One is personal and developmental, the other evolutionary. Both lead to stories that end with changes in our microbiome now being caused by technology and culture. Like the Earth as a whole, our personal microbial planet is subject to alterations we have brought about ourselves. This chapter sketches the current versions of both stories, and asks what we have done to the modern microbiome.

Bringing up baby

New babies are delicious, aren't they? Those tiny toes. That cute mouth. The miniature gestures. That smell!

Microbes think so, too. Here is a whole new collection of places to live, awaiting tenants. Whole plains of unoccupied skin. A new gut, getting busy with digestion, nicely furnished for adoption as a new home. To the parents it is a lovable miracle. To the microbes, more an unbeatable real estate opportunity.

And like most such opportunities, that way of describing it is too good to be true. It is not a free-for-all. Some microbes have the inside track on the deal, thanks to mother.

For most of history it went like this, we thought. An embryonic new human grows into a big-headed bulging-brained baby. When its head is almost, but not quite, too large to pass through its mother's bony pelvic girdle, loosened in the ligaments for the occasion, intense muscle contractions push it down ... and out. Along the way that gives it all-over close contact with vaginal microbes before it reaches the outside world. As labour often takes hours, there is plenty of time for fast-dividing bacteria to begin multiplying on the baby's skin before birth. And at the last moment, the still-soft skull exerts pressure on the anus and often squeezes out a little of the contents, anointing baby's head with maternal shit. Just before the challenge of taking his or her first breath, the newborn has a generous inoculation with a dose of healthy, human-nurtured microbes. Until then, it is microbe-free.

But can we be sure of that, when so many adult body regions once thought sterile have yielded signs of microbial life? Well, the claim stood for a long time. Baby's own first bowel movement, expelling tarry meconium, was declared sterile by Escherich, he of *E. coli* fame, when he looked into the matter in 1895. As recently as 2012 a review article in the normally authoritative scientific journal *Nature* began with the declaration that 'Every one of us enters the world devoid of microbial colonisation because of the sterile environment of the womb'. The assertion was unreferenced. In this scholarly context that is the convention for things the authors assume are generally accepted.[1]

It had, however, already been contradicted five years earlier, in 2007. First mice then, a year later, humans were found to have meconium already colonised by bacteria. They include *Lactobacilli* and good old *E. coli*. The first are common in the vagina, and increase in abundance during pregnancy. The others are gut bacteria. So how do they arrive in the meconium, which accumulates in the bowel before the first feed and is the residue of what baby swallows in the uterus – a mix of amniotic fluid, dead epithelial cells and other gunk? The mice in the first, Spanish, study that contested sterility definitely passed gut bacteria on to their babies somehow. The mothers drank milk containing bacteria with a distinctive genetic marker, and baby mice delivered by caesarean in operating-theatre cleanliness expelled the same bacteria immediately after birth. It's not clear yet how they get there.

Perhaps some bacteria are simply swallowed with amniotic fluid. However, yet another new finding is that the normal placenta also carries a bacterial load. A study from Kjersti Aagard of Baylor College of Medicine in Texas confirmed this by analysing placental tissue collected from 320 mothers after delivery, and duly reported a low but diverse bacterial population in the samples.[2]

Her paper also reported, and we need to be careful here, that the population mix resembled the microbiome of the mouth more than anywhere else. Cue news stories proposing that oral hygiene and dental health must be a priority for women who are pregnant, or trying to get pregnant. This, on the evidence, is going too far. Recovering the DNA signatures of particular bacteria from a tissue does not tell us how they got there. It is true that gum disease and premature birth sometimes go together, but both could be promoted by some other factors. So the benefits or costs of having some, or no bacteria in the placenta won't be clear before further close study (though it is still a good idea to clean your teeth often enough to help prevent gum disease).

This study also failed to show any link between bacteria in the placenta and the newborn babies' own microbiomes, so probably isn't the answer to the meconium microbe mystery either. At the moment, the best guess seems to be that immune cells in the mother's intestinal epithelium that can latch on to gut bacteria carry some of them into the bloodstream. That potentially offers a route by which they could be delivered anywhere else, including the placenta, the uterus and, yes, the foetus. Designing a study to test that idea, whether in people or lab mice, is probably taxing someone's ingenuity as I write. Until that is done, another line of speculation – that the foetal microbiome could be fine-tuned by giving mothers some sort of probiotic cocktail – is also likely to remain just that: speculation. Without a clear fix on the mode of transmission, it would be haphazard, at best.

After birth, babies enter a microbe-rich world and obviously acquire some of their very own microbes from the people around them – few of whom can resist cuddling, cooing over or nuzzling a neonate (maybe evolution helped us acquire a liking for the smell to ensure this happens). Equally obviously, once they can get finger to mouth, and move around a little, small humans can sample microbes from pretty much anywhere they can reach, including their own bodies, other people, pets, and all those things we tend to tell them not to touch, knowing we will probably be ignored even if we are understood.

The mother's own vaginal microbes alter before she reaches term, to help ensure the baby is inoculated with a useful mix of species. But there is another way she prepares to influence her child's microbiome: by producing breast milk. This, too, was long believed to be sterile unless the mother was ill. The excuse that the bacteria were hard to culture may apply, but one suspects that the idea that the suckling infant was guzzling down bacteria just didn't fit the idea of the healthfulness of nature's own baby food.

However, we now know that healthy breastfeeding mothers have hundreds of species of bacteria in their milk. They also manage to alter the make-up of the population to suit the baby's needs. Immediately after the baby is born, its diet includes a larger dose of lactic acid bacteria, as well as *Staphylococcus* and *Streptococcus*. Six months later, the mix is more like that normally found in the mouth, perhaps because babies usually begin to take some solid (well, mushy) food around then.

Here is another recent discovery that offers a puzzle about transmission. There are bacteria in breast milk as soon as it starts to flow – in fact they are found in the colostrum, which is ideally baby's first feed. How do they get there? We do not know for sure. The breast itself carries a microbiome, which contributes some species. There are bacteria on the skin around the nipple (of course there are). But some of the bacteria identified in breast milk are normally found in the gut, not on the skin, so the search is on again for an internal microbial highway.[3]

The most intriguing possibility involves the specialised cells called dendritic cells* that sample bacteria in the gut and

---

* More on these in Chapter 7. They are called dendritic cells because of their long, spidery extensions that can reach in between other cells such as those in the gut epithelium. These are not to be confused with the dendrites on nerve cells, which are the smallest elements of nerve fibres and look superficially similar.

present their characteristic antigens for consideration by the immune system. They can also migrate and actually carry bacteria inside the tissues, at least as far as the nearest lymph node, and perhaps the spleen. They could be candidates for purposeful microbial transport to other parts of the body. Experiments in mice have shown bacteria in lymph nodes make it to the mammary glands in late pregnancy. That is as far as the story goes at the moment. But it does look as if, by some such subtle means, the right bacteria must be selected, protected from the innate immune system, and ferried to the breast.

We know more about two other contributions the mother makes if she breastfeeds. Her milk contains plenty of immunoglobulins, which help stand in for baby's own gut immune system until it begins to develop. And it is formulated to provide microbial nutrition.

This is another discovery that is quite unexpected at first glance. It is one of the points where the developmental and evolutionary stories can't be cleanly separated. And it shows how closely interwoven our lives are with our microbes. As you would expect, human breast milk is stuffed full of easily digestible food to meet the growing baby's needs. It also contains chemicals that the baby cannot digest. Not, that is, without microbial assistance.

Specifically, a collection of complex carbohydrates known as human milk oligosaccharides are the third largest component of breast milk. (you may see them referred to as glycans, too). Once in the gut, they serve as ideal food for *Bifidobacterium infantis*, a bacterium that is not present at birth but soon normally becomes one of the biggest populations in a breastfed baby's microbiome. Making molecules like this is a costly business, metabolically speaking, and they are being churned out in quantity when the woman's biochemistry is working at full stretch to feed her baby. It turns out she is not 'eating for two', but for billions. Bruce German of the University of California, Davis sums it up: 'Mothers are literally recruiting another life form to baby-sit their babies and using the oligosaccharides to direct the microbiome.'[4]

All this ensures the infant intestines do not remain uncolonised for long. The maturation of the microbiome is then a matter of one population supplanting another, not microbes entering an unclaimed space. And we have a pretty good idea now what happens to the microbiota after the first populations are established. The most detailed work has been done on the populations of the gut, with this most complex ecosystem also seeing some of the largest changes.[5]

At first, the neonatal gut seems to host bacteria that can live in environments with or without oxygen. As they use up what little oxygen is there, the conditions change to favour those that can live without it. From now on, the population of the gut is restricted to anaerobic bacteria – this is one of the reasons many of the species eluded researchers trying to culture them in the lab, as anaerobes are generally harder to cultivate.

So after a couple of weeks the first population balance that lasts an appreciable time is usually established. It is heavily loaded with *Bifidobacterium*, *Bacteroides* and *Clostridium*. Although we did not appreciate quite how special some *Bifidobacteria*'s relationship with breast milk is until recently, their general role here is an ancient discovery by microbiological standards. They were identified in samples from breastfed infants at the very end of the 19th century.

When a baby starts to eat other food, normally after a few months, the gut bacteria gradually diversify. Those who have been bottle-fed have more diverse species earlier on, and the difference is lasting – they are likely to have a population balance typical of adults earlier on than their breastfed contemporaries.

Caesarean delivery, on the other hand, begins a different course. The first colonisers are typically skin microbes, from the mother and anyone else who handles the baby. Others soon follow, but sterile delivery, almost invariably accompanied by antibiotics for the mother before birth, leads to less diversity in the gut microbiome, as does premature birth.

As babies begin to explore their world, they are also exploring the local microbial flora, and random encounters begin to play a larger role in which species are trying their luck living on this promising new creature. Infants' gut microbiome samples (conveniently provided as often as you like, or even more often) vary more than those from adults, both in composition and in those functional genetic modules that make adults look more similar than their species analysis. There are obviously a host of variables that affect how this goes in individual households. One of the most influential is that humans may not be the only superorganism around. A paper from University of Colorado DNA specialist Rob Knight's lab in 2013 tells you pretty much everything you need to know about the reported findings in the title: 'Cohabiting family members share microbiota with one another and with their dogs'.[6] Admittedly, the admixture was more thorough for skin microbes than those found elsewhere, but the study, which covered 60 families, of which 36 kept a dog, did show clear differences between the human microbiomes of dog-lovers and the rest. Cats, for all their strokeability, made a smaller contribution to the creation of a communal, rather than individual, microbiome in a household.

Environment and circumstance must play a part in shaping

a baby's microbiome. However, active selection may still turn out to be more important. A report from Washington University School of Medicine in 2014 suggested that all premature babies' microbiomes develop much the same way. They followed 58 premature infants who started life in a neonatal intensive care unit, where unplanned microbial encounters are rare. The way they were delivered, and how they were fed, made little or no difference to development of the gut microbiome, which went through 'a choreographed succession of bacterial classes'. There were differences in the speed of changes, but not in their sequence, and by 33–36 weeks after conception all of these babies had very similar microbial populations in their guts.[7]

In any case, as the baby eats more, and more varied, solid food, and is eventually weaned off human or formula milk, any differences that are the legacy of delivery or early feeding gradually even out. Three-year-olds have a microbiome that looks pretty much the same as that of an adult who shares their diet and environment. They have acquired a large, mixed population of *Bacteroides* and *Firmicutes*, though some *Bifidobacteria* remain. And that is how the ecosystem most often remains, unless destabilised by some out-of-the-ordinary event.

As snapshots are supplemented by more studies that follow one or more people over time we get a better idea of what kind of events these might be. The human microbiome lends itself easily to self-study by the dedicated, and one particularly detailed example comes from a study by Eric Alm's lab at MIT. They followed two people who sampled their stool and saliva every day, and used a handy iPhone app to record what they ate and drank that day, along with a ream of other relevant data about work, sleep, exercise, mood and body weight. The idea was to use the accessibility of DNA sequencing to track microbial population shifts from day to day.

The two people tracked were not named in the formal paper published in mid-2014, which in the conventionally coy language of science referred to them as subjects A and B. 'Our screening yielded a small cohort of two healthy, unrelated male volunteers,' the authors reported.[8] They were in fact Alm himself and another researcher, Lawrence David, who when the study began was a graduate student in the lab. This one was personal. It might not have been, but when the plan was first trailed, with the demand to take a sample every time the volunteer visited the lavatory for a year, 'surprisingly, no one signed up for the study,' Alm admitted.

The results are an ecologically fascinating blend of variation with continuity. In this sample of two, you can always tell which person is which, even though both vary from day to day. The overall picture is one of stability.

Day-to-day variation was partly linked to meals. Eating larger-than-usual amounts of fibre led to a rapid increase in three bacterial species, measurable in the next day or so, though they were species that in total accounted for only a small proportion – around a fifth – of the overall population. Citrus fruit boosted a different cluster of species.

Another influence that showed up was serendipitous. David spent two months in Bangkok in the middle of the year, and the species abundance plots show clearly that some bacteria not present previously take up residence in his gut as soon as he arrives, and remain until he returns to the USA, when they disappear again.

A more dramatic event marked the microbial record for David's stay-at-home colleague. A bout of food poisoning from a restaurant meal saw an uncomfortable demonstration of the power of bacterial reproduction. The average presence of *Enterobacteriacaea* – the parent family of the food poisoning bug *Salmonella* – in Alm's gut over the year was 0.0004 per cent. When he was feeling ill and suffering diarrhoea it went up to 10 per cent of the total, then peaked at an impressive 29 per cent. At the same time, his *Firmicutes* were severely reduced. When he got better, they went back up to the 40 per cent level he was used to, but the population mix of strains living in his colon, in a newly stable state, were different from before. Most of the newly abundant kinds, though, had been present in small amounts before he got sick, so the re-establishment of the population did not depend on complete newcomers. It is another indication that a particular person's microbiome likes to maintain a similar mix of microbes over time. It can happily cope with substitutions of species as long as the replacements do more or less the same thing, and there are bacterial resources on hand that can quickly fill a niche in the ecosystem when it gets depleted.

The picture that emerges from this one small study, then, is one of resilience, to use a term that is often applied now to healthy people's microbiomes. These two personal microbiomes were different, but stable. There was evidence of microbial competition in both collections of species, with strains that were closely related sometimes changing in abundance – one increasing, one decreasing – suggesting that one had gained a competitive edge. But these minor day-to-day shifts in the ecology did not affect the overall picture much.

Other studies, such as one from Jeffrey Gordon's lab in 2013, have added to the evidence that the gut microbiome's composition tends to be pretty stable, in this case in samples taken from three dozen adults over five years. For many people, it seems, once the adult gut microbiome is established, it stays broadly the same, barring illness.

That stability is also more likely to be disturbed as people age. Studies of the gut microbiome in the elderly suggest there is more variation between them than in younger adults, while each individual may also have a less diverse microbial population than that typically found in samples from young adults or the middle-aged.[9] The details of these changes are still being worked out, with efforts to disentangle lots of differences that may each affect any one person's microbial story – elderly people in developed countries may be in care homes, eating different diets from their previous habits, and take a daily cocktail of prescribed medication. The microbiome goes on changing in response until death. Then the microbes will do their best to carry on elsewhere.

First, though, they will consume the nutrients that leak from our dying cells. That, too can be studied – the microbial ensemble living on a corpse even has a proper scientific name: the thanatomicrobiome. Peter Noble of Alabama State University suggests it may have forensic uses, such as estimation of time of death.[10] At that point, though, there is no longer a superorganism, so it is time to move on to the larger history of microbiomes.

Evolving together

Underlying the individual story of what shapes each person's gut microbiome is a much longer, evolutionary story that relates how we came to be sharing our lives with all these species. The new technologies are helping to piece this together, too, partly by dissecting the microbiomes of lots and lots of other creatures. Call humans a superorganism and it sounds as if we are giving ourselves airs. But look around and it is immediately apparent that almost everything else is a superorganism, too. Were they always 'super', or did they acquire this quality gradually?

We know that bacteria were here first, with an evolutionary head start of a couple of billion years. So the real question is how they were involved in the evolution of more complex organisms. They were a crucial part of the origin of eukaryotic cells, through the bacterial ancestors of mitochondria. But what happened after that as the newly fledged eukaryotes made their way in a bacterial world?

To begin with, the interactions were probably simple. Take a look at another single cell. This one has a nucleus, so it is a eukaryote like us. It is roughly egg-shaped, and has a tail, or more properly a single, movable, whip-like flagellum, at one end. Look a little closer and the base of the flagellum is surrounded by a collar of a few dozen microvilli.

Its flagellum moves the beast through the water, and the collar traps things carried on the current, including bacteria. Our cell then eats them. This is a collared flagellate, or choanoflagellate, one of the simplest eukaryotes. There are no fossil remains, but the archive preserved in its genome suggests that it assumed something like its modern form 800 or 900 million years ago. The creature that lived then, DNA evidence indicates, was the common ancestor of two lineages: present-day choanoflagellates; and the first animals.

We now think they were already interacting with bacteria back then in more subtle ways than treating them as prey. One present-day species of choanoflagellates responds to a specific bacterial signal by forming colonies. Could this mean bacteria were involved in the origins of multicellular organisms? Perhaps it could. It certainly suggests that our very early animal predecessors were already registering chemical messages from bacteria.

The point is underscored by another observation that gives pause. A colony of choanoflagellates looks remarkably like a sponge. And marine sponges, the simplest of contemporary multicellular organisms, are often full of symbiotic microbes, amounting to as much as a third of their cell mass. As sponges, like choanoflagellates, also eat bacteria, there must be a mutual recognition that allows some of them to cohabit with the sponge cells instead of being swallowed up with their bacterial brethren.

This furnishes strong circumstantial evidence that bacteria were closely involved with the lives of multicellular organisms from their first appearance. Biologists did not think much about this kind of thing until the 1980s. Since then, evolutionary relationships have become clearer in the wake of Carl Woese's work on bacterial classification. And the enormous diversity of bacteria everywhere you look, revealed by mass DNA sequencing, has induced researchers to pay more attention to how microbial life interacts with everything else.

The importance of co-evolution of *all* the later emerging species with bacteria comes across from the sheer range of microbiomes that have now been studied. Aphids and ants, butterflies and moths, lice and fruit flies all frequently carry symbiotic bacteria, and we understand some of the essential things they do for their hosts.

To give just one example, one of the most extraordinary inter-species relationships is demonstrated by leaf-cutter ants, which harvest leaves in tropical forests and use the partially chewed leaves to grow fungi, which they can then eat – an arrangement to marvel at, in which a colony of ants millions strong can process hundreds of kilos of vegetation every year in an underground fungus farm. In 2010, it was shown that the fungus chamber breaks down the leaves with help from a community of microbes resembling those found in the cow's rumen, the extra stomach where the larger herbivore digests grass. Two solutions to the same problem – extracting nutrients from tough plant material – have evolved, the cows maintaining a microbiome that does the job internally, the ants externally.*

The same paper refers in turn to studies of antibiotic-producing bacteria in other species including insects, plants, corals, sponges, snails and birds (the hoopoe, since you ask). In short, everywhere you look, there are microbiomes tailored to the job that the small organisms are doing for the larger one. The 'eat or be eaten' side of evolution can lead to oddities in the relations between them. Termites have in their hindgut an unusually complex microbiome for an insect, with a mix of single-celled eukaryotes and bacteria. The microbes' digestive aid allows the termites to basically live on wood. A range

---

* This particular marvel of co-evolution goes further. The ants also have a microbiome of their own on their tough exoskeleton, which supports bacterial species that play another essential role in this system: they make antibiotics that keep down pathogens that would otherwise attack the fungus.The ants ensure a rich nutrient supply, and this favours bacterial species that compete with others by direct interference – meaning antibiotics – rather than speed of growth. The beauty of the set-up is that the ant does not know it is selecting bacterial species with a particular property – and no specific signalling or recognition is involved. It has just evolved to create conditions where self-selection happens as the various bacteria do what comes naturally. See Scheuring, (2012).

of larger creatures, including armadillos and aardvarks, have microbiomes in their guts specialised to allow them to eat chitinous ants, and termites. It is microbiome against microbiome out there.

This larger appreciation of the importance of bacteria was turned into a manifesto in 2013 by Margaret McFall-Ngai of the University of Wisconsin, whose ideas about the immune system will enlighten us in the next chapter. She recruited two dozen other influential biologists to help draft a call to arms for researchers that was published in the *Proceedings of the National Academy of Sciences* under the heading 'Animals in a Bacterial World'.

The strands they pull together mainly come from study of animals alive now. McFall-Ngai and her colleagues have spent decades studying the symbiosis between a tiny squid and the *vibrio* bacteria that live in a luminescent organ inside its body cavity. Such present-day interactions furnish lots of hints of how such things went in the past.

An unexpected finding from her favourite model system is one of the best examples. It involves two molecules: the lipopolysaccharide (LPS) that sits on the outside of *vibrio*, and the peptidoglycan it uses to strengthen its cell walls. In large multicellular organisms we know that these two molecules can activate the immune system. But McFall-Ngai found that the *vibrio* also uses them in an entirely different way. When they are present they induce development in the squid tissues that will accommodate the bacteria.

It then emerged that the same two kinds of molecules (found on the surface of bacteria, remember), need to be present to ensure normal development of the gut in mice. So a signalling pathway that begins with bacteria has been preserved over a long stretch of evolutionary time, and – as often happens – used for new things later on.

McFall-Ngai has a knack for combining minutely detailed work on the intricacies of one organism with large-scale speculation about the overall evolutionary course of events. In particular, she argues that we can piece together a series of crucial stages in evolution of bacterial–eukaryote relations that add up to a plausible continuous narrative of this crucial aspect of evolution.

It goes like this. All the original lineages of complex organisms began in the oceans, which surrounded them with dissolved organic matter and an ever-present population of bacteria – between 100,000 and 1 million in every millilitre of seawater. A sponge shows the likely outcome if you are a multicellular organism with no tissues. You simply end up with bacteria all over the place.

Once multicellular organisms move beyond the level of organisation of a sponge, they have a body wall. It is an easy option when you are in the ocean to take in nutrients across that wall, where our skin would be, which begins to look like the absorptive cell layer we now find in places like the intestinal epithelium. Its other function is as a barrier, preventing bacteria making use of the nutrients that our new multicellular creature is busy extracting from seawater and concentrating for its own use. As McFall-Ngai put it in a talk, 'they're sharing the nutrient pool, and at the same time trying to stop these organisms from growing all over them'.

A rather more developed creature than a sponge is the freshwater polyp *Hydra*, a relative of coral, jellyfish and sea anemones. Its organisation is relatively simple, but it still has a body plan – basically a tube with a mouth at one end and a 'foot' that anchors the tube at the other. The tube is surrounded by two cell layers. The outer layer is epidermal, the inner is now gut epithelium. *Hydra*, in its few millimetres of body, has a stomach, and the important business of absorbing nutrients has now (at least 500 million years ago) been brought inside. The whole creature is basically an early model intestine. And *Hydra* already has a simple microbiome. Thomas Bosch of Kiel University and colleagues have shown that different species of modern-day *Hydra* have distinctive microbial populations, too. In other words, it is selecting the bacteria it chooses to live with. It has also developed strong defences against the ones it does not favour, including producing antimicrobial peptides and enzyme inhibitors. The microbes it does tolerate, or even encourage, are already deeply woven into the life of this small creature. *Hydra* are famous for their capacity to reproduce asexually, by budding. But if they do not carry their normal microbiota, this ability disappears. Inoculate the microbe-free *Hydra* with cultures from the guts of normal specimens, and budding is restored.[11]

Once on land, animals have no choice but to adopt this outside-in arrangement, with the nutrient-absorbing tissue inside the body. Then we get many versions of the gut and eventually, in vertebrates, a more complex microbiome in the gut. There is a conspicuous difference between the multicellular creatures that have relatively simple resident microbiomes, like insects, and ones like us that usually have much more complex microbial communities. The latter, all vertebrates starting with the fishes, have an adaptive immune system in addition to the simpler, and earlier evolved, innate immune system. McFall-Ngai suggests that the adaptive system, with its complex of antibodies and many kinds of lymphocytes, may have evolved in order to manage co-existence with a complex microbiome. Once again, there is a pointer to contemplating the immune system as the key to the way our superorganism works.

Meantime, there are other, less complex developments that may be part of the story of host–microbial co-evolution. The gut wall separated from the body. This allowed the intestines to loop and coil inside the body cavity, and guts grew longer, especially in herbivores. They could swallow more food, which took longer to be processed and could be dealt with in stages in a gut that has distinct regions with different chemistry. Could bacteria have encouraged all this by acting on hard-to-digest food molecules during their slow passage from mouth to anus? Perhaps they could.*

Another curious fact the manifesto highlights is that most of the bacteria that live in the guts of birds and mammals grow best at around 40°C. That might just be because these warm-blooded creatures happen to operate at that temperature and proved hospitable to bacteria that like to do the same. But just maybe it is the other way round. There was energy to be gained by recruiting bacteria to help digest those complex carbohydrates, and they do it better if you keep them warm. Perhaps we have our gut bacteria to thank for the fact that we are warm-blooded!

But for now, let us jump a few hundred million years to creatures like us that already have complex gut microbiomes. Whatever happened in earlier evolutionary eras, we can see

---

* I concede that size is relative, so the guts of, say, a fruit fly, are not imposing relative to any mammal. Still, they are larger than the 20 or so species of bacteria that dwell there. Anyway, I rather like this particular speculation and look forward to seeing if anyone can think of a way to try and test it.

that a complex gut microbiome is something we share with other primates.

Researchers have sampled bacteria from quite a few other primates' faeces by now: captive and wild chimps and gorillas, baboons, colobus and howler monkeys, and more. The results indicate that they, like us, have species-specific, stable microbiomes as adults. These can be related to their diets – most of them eat a *lot* of leaves but the dominant plants vary – and are susceptible to outside influences but the selection each species makes remains identifiably its own. Black howler monkeys, for example, have more diverse microbial species in their gut if they live in the rainforest than in the less richly varied vegetation of a temperate deciduous forest.[12]

Humans came out of the same lineage as the ancestors of these contemporary apes and monkeys, but we would expect them to have different gut microbiomes. We are omnivores, and have been using fire for perhaps a million years. Cooking is basically a technology for starting digestion before you eat the food, and entire theories of human evolution have been developed based on the idea that it allowed us to develop larger brains by providing more efficient access to high-energy protein, as well as reducing the time hunter-gatherers have to spend chewing their dinner. But leaving that intriguing suggestion aside,[13] cooking alone would probably have led to a different gut microbiome from *Homo erectus* onwards.

How different? We cannot go back nearly as far as *Homo erectus*, our upright proto-human ancestor of 800,000 years ago, but some heroic DNA analysis has produced evidence of the microbial content of the remains of ancient human stool.

The samples, known as coprolites, that were reported in a paper from Cecil Lewis at the University of Oklahoma in

2012 came from three different sites, the oldest of them a cave in the southern USA that was inhabited 8,000 years ago. The others were between 1,000 and 2,000 years old. Two of them were, frankly, hard to interpret, unsurprisingly for such old material. But one of the more recent samples, from a site in northern Mexico, was closer to gut samples taken from child country-dwellers in present-day Africa than to the city-dwellers in the USA who furnished samples for the Human Microbiome Project.[14]

This is the kind of finding that points to changes in the human gut microbiome since *Homo sapiens* first appeared. A lot hinges on this, as there are many hypotheses about how departures from the ancestral human microbiome could affect us. The story I told earlier about the disappearance of *C. difficile* from the majority of infant stomachs is one example. There are many other, often more speculative, ideas about links between altered gut microbiomes and health. But what kind of alterations are we talking about?

A thoroughly modern microbiome

The human microbiome as we expose it to scientists' informed gaze today has been shaped by more than evolutionary pressures. We are inventive creatures that modified our environment to suit ourselves. That has had effects on the microbial portion of our superorganism we need to know about. Start with our eating habits.

We know that the balance of microbial populations responds to changes in diet from day to day, and there is good evidence that humans' eating habits over much longer time spans have effects too. The longest convincing look back comes not from fossilised faeces, but from teeth.

We know in principle that people and bacteria have always co-existed, but finding traces of the bacteria that lived on us in the past is a tall order. They mostly disappear after death, along with the soft tissues.

The exception, apart from fossil poo, is one that would have gladdened Leeuwenhoek's heart − preserved dental plaque.

Biofilms, those complicated communities of bacteria that co-operate to coat surfaces that give them convenient access to nutrients, form easily on teeth. The film that grows over the surface of a tooth starts soft, but gradually turns into tartar. This is pretty much like concrete. It turns out it can preserve not just food particles but also dead microorganisms, or at least some of their molecules, including DNA.

Getting the information out again is no mean feat. Analysing the microbiomes of the living is tricky, but think about working with dead people's teeth. Archaeological specimens may well have been handled by researchers decades ago who had never heard of the microbiome, so are prone to contamination. And the amounts of plaque recovered are small. Still, if you are really, really careful, you can get enough bacterial DNA to run 16S rRNA analysis.

Do that to a suitable sample of ancient skeletons and you can open a window on to the co-evolution of people and our microbiota − the oral rather than gut microbes this time. Analysis like this shows up two big changes. The first, perhaps 10,000 years ago, coincides with the origin of agriculture. The second, much later, marks the era of processed food that shapes so many people's diets today.

An international team led by Christina Adler of the Australian Centre for Ancient DNA reported in 2012 that they had been able to analyse samples from 34 prehistoric human skeletons.[15] Eleven were male, eleven female and the remaining dozen could be either – 'skeleton' here can just mean small fragments dug up after thousands of years. They covered a span of human history from before the agricultural revolution, represented by some of the last hunter-gatherers in Poland, who lived there a bit less than 8,000 years ago, up to late medieval people, a few hundred years ago.

The overall picture was not dramatically different from the bacterial populations of present-day mouths. All fifteen of the main groups of species that have been found on 21st-century teeth and gums were already happily ensconced in human mouths thousands of years ago.

But the detailed inventory shows two important changes since our ancestors' time. Inspection of remains had already established that early hunter-gatherer groups had less dental disease (tooth decay and gum problems) than farming peoples. The microbiome signatures show why this is, with bacteria associated with tooth decay (*Veillonelaceae*) and gum disease (*Porphyromonas gingivalis*) appearing in the later samples. An increase in carbohydrates in the diet after the agricultural revolution could account for their appearance. Actual tooth damage tended to appear later in life, but the offending bacteria were found in plaque samples from the youngest child in the study, who was estimated to be only three or four years old.

A second shift occurred on the way to the modern oral microbiome, which typically harbours more of the decay-causing species and one in particular which is strongly associated with dental caries – *Streptococcus mutans*. This also fits with what we already know about diet. Agriculturalists' eating habits stayed much the same until the Industrial Revolution, when people in developed countries began eating more refined grains and enjoyed a new world of sweetness from concentrated sugar extracted from sugar beet and cane. That tasted good, but left mouths dosed with sugars that bacteria can ferment into acids. As they are made directly in the plaque it is easy for them to eat into the tooth underneath, dissolving the mineral matrix of the tooth and creating a cavity. Welcome to the modern world.

This shift toward the pathological goes along with a reduction in the overall diversity of the oral population. The researchers reach for the analogy with what we think we know about larger-scale ecosystems here, and suggest that this may make the oral microbiome of sweet-toothed moderns less resilient, and more likely to be disturbed by dietary imbalances (they don't say what kind) or 'invasion' by nastier species.

It is, in short, (yet) another version of the fall. Once, we were happy hunter-gatherers, with a diverse, ecologically sound biofilm setting the seal on nice healthy teeth. Now, we have known sin, or at least sugar, and our mouths bear the mark.

A similar story can be told about the gut microbiome, partly from the rather slender evidence from coprolites, partly from differences in results from studies in different parts of the world. We cannot really know what the earliest human gut microbiomes looked like, but we can investigate the correlates of differences in lifestyle today.

As DNA typing of microbial populations in the gut has extended into more populations, it looks as if there are significant differences between, say, the urban US populations studied in the Human Microbiome Project and some other groups.

An early indication here was a 2010 study comparing children in a rural village in Burkina Faso and a group from Europe.[16] The African, agrarian community had a diet with more plant protein, carbohydrate and fibre than the Europeans. That went with greater microbial diversity, an overall lower proportion of *Bacteroides* in the gut, and a higher production of the energy-boosting short-chain fatty acids. A subsequent comparison between gut microbiomes in the USA, Malawi and rural Venezuela, which looked at both children and adults in all three countries, produced rather similar results. Again, the population mix in the guts of people living in the USA was distinctive.

More refined comparisons continue to appear, the latest as I write reporting on a study of the Hadza hunter-gatherers who live in Tanzania.[17] They are one of the last traditional hunter-gatherer groups anywhere, with 200–300 people living close to the way we imagine everyone lived before settled agricultural communities were invented: around 90 per cent of their food still comes from hunting or foraging. The supposition here is obviously that they might be closer to an original human microbiome, if there were such a thing. There were the now-expected differences between this group and a control group of city-dwellers in Italy, a group who ate what is considered a healthy diet by modern standards, with lots of fruit, vegetables, olive oil and pasta, but who still consumed plenty of sugar and easy-to-digest starch. The additional finding this time was a further set of differences with the gut microbiomes of the already studied rural farmers in Burkina Faso and Malawi. The hunter-gatherers had gut populations enriched in populations thought to help digest large amounts of fibrous plants.

A separate study of an isolated group of Amazonian hunter-gatherers, led by Maria Dominguez-Bello of New York University, is under way and may have published results by the time you read this. Early hints are that it also reveals a gut microbiome with greater diversity than that of, say, a typical person in Houston, Texas.[18]

So it looks very plausible that we have simultaneously made two large discoveries: that the gut microbiome, especially, is complex, diverse, and important; and that it has changed under the influence of modern life. Interpret that with caution, though. Broader studies of ape and monkey species suggest that all humans, no matter how or where they live, have less diverse gut microbiomes than all their evolutionary relatives among the primates.[19] Something, perhaps shifting to eating more meat as people got better at hunting, perhaps using fire to cook it, began changing the human microbiome long before modern culture began to influence the life in our guts. We need to understand more about that shift before reaching a verdict on the more recent changes.

Those changes, though, do look significant, and go well beyond the advent of processed food and lack of fibre. Antibiotics save lives, but at the cost of severely depleting the microbiome. It usually recovers, but the effects of repeated use, especially in children, are more long lasting. Add the increasing proportion of births by caesarean section in many countries (now around one in three in the USA), and our liking for disinfectants and cleansers, and many people now clearly live in a very different microbial environment from the one humans experienced for most of their history.

This must affect our microbiota. The common view is that the gut microbiome, in particular, tends to be less diverse and less stable nowadays. This is associated, in many minds, with the modern increase in a whole range of medical problems. In its strongest form, as outlined by Martin Blaser for example, we face a 'disappearing microbiota' which is 'exacting a terrible price'.

That seems an exaggeration for the microbiota as a whole. It is not really disappearing. But there are some less often remarked modern advances that have definitely led to disappearance of some components of our inner menagerie. It is estimated, for example, that as recently as 1940 perhaps 70 per cent of children in some rural parts of the USA picked up infestations with the tiny parasitic worms known as helminths. These creatures – roundworms are the most common example – are still very common in many parts of the world, but now rare in the West.

But it is clear that the microbiota proper *are* changing. We do need to know if that matters. It is tempting to link the change to conditions that seem to be getting more common, such as allergies or autism. We need to be sure this is thought through properly, though, and not just based on an assumption that 'original = natural = good'. There is no reason to think an altered microbiome is necessarily a bad thing in itself. If the evolution of our gut microbiome was driven by the advantage it confers in digesting more and different foodstuffs, the complementary advantage is that the immense diversity of microbial genes makes it very versatile, and their swift reproduction allows them to adapt rapidly to changes in the diet. The ease with which they shift population is part of the point of microbes being there in the first place.

Still, the list of problems that have been linked to alterations in the microbiome – as cause or consequence – is long, and I will review them in Chapter 8. Making sense of the connections will be easier, though, if we first consider the aspect of this that I have missed out of the evolutionary and developmental story. When you investigate the research on our personal microbes and virtually any condition, from allergy to depression to heart disease, sooner or later the discussion brings in one other crucial feature of our superorganism: the immune system. Here, if anywhere, lie many of the keys to making sense of what our microbiome does for us, or fails to. But can we locate them?

7 | Working together

A cell biologist talks to an immunologist: 'I did this experiment and IL6 [interleukin 6, a signalling molecule] went
up. That's bad isn't it?'

Immunologist: 'Yes, it's bad.'

A week later – 'I changed the experiment and IL6 went
down – so that's good, right?'

Immunologist: 'No, no. That's bad as well.'

Well, it made a microbiome conference laugh. This small
joke expresses two things about the tribal lore of different bands
of biologists. Immunologists are seen as bad at communicating
their stuff to everyone else. And even if they try, it may be hard
to understand what is going on – it doesn't necessarily make
sense to them, either.

But there is no getting round having to deal with the
immune system. The human microbiome interacts with all
four of the great information and control systems of the body
– the genome, the endocrine system of hormones, the brain
and nervous system, and the immune system – just as they all
interact with one another. This is one reason why investigating
the microbiome is driving new, interdisciplinary conversations.

The signal traffic between the microbiome and the immune

system, though, is probably the most complex. It has origins deep in evolutionary history. It is very likely that neither would be there, or not in the same shape, without the other. As biologists work together to picture what it means to be a superorganism, the ways we think about our immune system are being transformed. As well as being a vital part of the microbiome's normal diplomatic negotiations with its host, communication – or miscommunication – with the immune system is part of almost every mechanism that has been suggested to account for links between the microbiota and disease.

The big rethink starts by rejecting the idea that guides almost all of our casual talk about our immunity. It is an idea that is another legacy of the germ theory.

The war within

There's one thing almost everyone knows about the immune system. Western medicine is a war against disease, and the immune response is the first line of defence. It helps us fight invasion by killer bugs. The world of molecules and cells becomes a battlefield, complete with alerts, invaders, recruitment and mobilisation, search-and-destroy missions and natural born killers. As a much-praised 1990s children's book tells it, 'This is the true story of the amazing defenders of your body, a heroic band of cells that keep you fit and healthy by constantly battling against all kinds of invader germs. Every second, every minute, every hour, every day of your life, they are fighting ...'[1]

This is more than a popular fancy. Medical school textbooks and immunologists' research papers use the same ideas. They are so central to the discipline that they often pass unnoticed. The war metaphor was immunology's trademark. The other main idea in play, that the immune system works by distinguishing between cells or antigens that are marked as 'self' or 'non-self', fit well with this notion of inescapable conflict. That is how you know who to fight.[2] The anthropologist Emily Martin, who spent time with immunologists in the 1980s, found that even those who felt uneasy about military metaphors could not see how to avoid them in their professional thinking. As one saw it then, 'no reasonable discussion takes place in the immune system'. However, he comforted himself with the thought that it was 'a just war'.[3]

The workings of the whole system, in all its cellular intricacy, threw up lots of findings that did not quite fit this framework, but the substitutes on offer did not convince that they worked any better. The idea that immune responses are triggered in response to injury or insult, registered as a danger signal, had a brief vogue in the 1990s but did not really catch on.[4]

But now, rather suddenly, there is a big change in the way scientists and doctors are thinking about the immune system. It is being called a 'paradigm shift', or a 'revolution'[5] and it was prompted by our recognition that an enormous number of cells that seem to be 'other' rather than 'self' are ever-present in and on our bodies. Far from waging war on them, we go to a lot of trouble to encourage them to take up residence. They must have a quite different relationship with the immune system than the one implied by it being primed only for defence against invasion. But what is it? The answers require a closer look at what this system consists of.

A brief guide to the immune system

If the immune system were a single organ, it would be a pretty large one. But it is not. There are some key locations for immune activity – those obscure organs the thymus and the spleen, bone marrow, lymph nodes, and so on. But the system is not located anywhere in particular. It has to have a presence everywhere in the body to do its job. There are immune molecules in blood, sweat and tears. The rise of microbiome science has put a spotlight on two other very important sites for immune system action: the skin and, above all, the gut, which perhaps unexpectedly is the largest immune organ of all.

What, then, is acting? Lots of things, is the unhelpful first answer. The immune system is ancient, like the bacteria. As we've seen, the first multicellular organisms that were getting organised to operate as individuals were surrounded by other, single-celled life, and needed ways of identifying which of the cells in the neighbourhood were part of the club, and which were interlopers.[6]

Faced with that basic distinction, our immune systems have evolved numerous mechanisms for identification, recognition and response over the aeons. Our current version is pretty complex. Part of that complexity arises from the necessity to recognise lots and lots of different kinds of foreign material through the action of receptors that are just the right shape – although we now know that this requirement is met with brilliant simplicity by having a small set of genes governing the structure of parts of antibody molecules which are 'hypervariable'. By swapping portions of DNA, a small number of genes can generate an effectively unlimited range of protein shapes, and thus provide a complementary match for any chemical shape that happens to turn up. Thus was the problem long known as the 'generator of diversity' (GOD for short) solved.

Many of the other complications are due to different ways in which a range of cells keep tabs on their surroundings, the ways they respond to signals when needed, and – just as important – the ways these responses are regulated.

We don't need a full inventory of immune gadgetry to talk about how the microbiome relates to the immune system. That would easily fill another book. But it is helpful to have a sketch of the main components.

The whole assembly of cells, signals and responses divides into two big categories. The innate immune system is a relatively simple recognition-and-response apparatus that you are born with. It is the most ancient part of the whole set-up, but was discovered surprisingly recently, in the late 1990s. In essence, it makes cells which can detect the presence of bacteria from characteristic molecules that the smaller microbes expose on their surfaces. That involves receptors for molecules such as the lipopolysaccharides that some bacteria carry on their outer membranes. The immune cells that sport these receptors are usually equipped to make antimicrobial chemicals. This does look like the defence system we were told about. Spot the dangerous invaders, and zap them! The vast majority of multicellular organisms only have an immune apparatus like this, and it seems to work well enough. Interestingly, with very few exceptions (mainly termites), they also have simple microbiomes.

The rest of our own immunity is kept up by the other limb, the adaptive immune system. That came later in evolution but contains the elements that biologists discovered decades before they knew about its older complement. It responds to less common challenges, basically by maintaining a vast repertoire of antibody-producing cells, any one of which can be provoked into fast replication if its unique antibody encounters a match. Most don't, but those that do go into a complex sequence of development that aids elimination of the matching antigen and maintains an increased level of the same antibody in future as an *aide memoire*. This is why vaccination works, for example: it primes the adaptive immune system. We put a lot of resources into building this immunological memory-and-recognition system. The number of immunoglobulin molecules in a millilitre of normal blood is 1,000 times more than the number of human cells in the entire body (10,000 trillion versus 10 trillion)[7].

The two arms of the system, innate and adaptive, muster a whole brigade of cell types. The main ones go like this.

There are sentinel cells, mainly serving the innate immune system, in many tissues. They include dendritic cells, macrophages and mast cells.

Next come circulating cells, in the lymph and the bloodstream and, when the action heats up, in various intracellular spaces. The ones to focus on are the lymphocytes, on which the adaptive immune system depends. They recognise particular antigens – chemical shapes – hence often sense the presence of a particular organism.

Lymphocytes also come in a vast range of types, with different roles. Things start simply, with a common set of stem cells in bone marrow, from which all the different kinds of white blood cell, including lymphocytes, are derived. Lymphocytes divide into two main populations. Some stay in the bone marrow as they develop and ultimately make specific antibodies. These are B cells. The others are T cells, named for the thymus where they migrate to finish developing. The thymus is a sort of further education college for immature T cells. There they complete their development. On graduation they are either 'effector T cells', which directly or indirectly attack infected cells, or 'regulatory T cells', which as the name suggests influence the action of all the other components of the immune system. Yet again, there are a *lot* of different types. A lymphocyte can have hundreds or thousands of different kinds of receptors on its surface, and T cell types tend to be named for the presence of one or more of the important ones (or sometimes, with a minus sign, for its absence).

Other crucial categorisations include double-negative or double-positive cells. The former refers to immature thymocytes that, very early on, fail to express two important cell markers known as CD4 and CD8 (hence 'double-negative'). More mature thymocytes that have started making both CD4 and CD8 are, logically enough, double-positive. As these cells become fully operational, as it were, in the immune system, they keep making either CD8 (mature 'killer cells') or CD4 ('helper cells', which orchestrate the immune response).

There are more layers to unpeel, but that is as deep as we need to go for now.

These varied cell types interact in elaborate signalling networks. Dendritic cells, for example, release signalling molecules known as cytokines when they sense the presence of an infectious agent or debris from damaged tissue. These act on nearby cells and blood vessels and summon other immune cells, which accumulate at the site in question. Blood vessel walls and epithelial layers loosen the junctions that usually seal the gaps between cells, allowing fluid to pass across the barrier and carry other immune components like antibodies. Together, these things add up to inflammation.

This is the main effect of the immune response that we notice. It is why the infected finger you pricked while gardening swells, reddens and feels tender (good). The pain you feel is not because of the infection, but comes from the inflammation.

It is also why the same finger reddens, swells and feels tender when you have chilblains (not so good). And just as we take some comfort from inflammation in some circumstances and find it a nuisance in others in these relatively minor cases, there is a highly elaborated set of evolved checks and balances in the system to try to ensure that inflammatory responses occur when, and only when, they will do some good. One measure of the complexity of this whole system is that we have catalogued around 50 different human cytokines so far. Most immune cells, and many other cell types, respond to more than one. Some prod immune cells into life while others signal them to carry on maintaining a watching brief, so the final outcome is often delicately balanced. Immune cells doing their stuff in the wrong place are at the root of many of the problems that multicellular flesh is heir to.

Even a simplified account of the immune system starts to raise doubts about 'cell wars'-style explanations. At first sight, the adaptive immune system may seem a useful addition to the body's armed forces. The cells that make specific antibodies to new foreign molecules refine the detection and surveillance capability of the defences and direct weaponry, in chemical and cellular form, to the right targets.

Look deeper into what goes on, though, and the idea that this is a military operation is a poor fit. Unanswered questions pile up. Why, for instance, does human immunoglobulin A latch on to surface molecules on some gut bacteria in a way that makes it easier for them cling to the gut wall and form biofilms?[8] And how and why did the adaptive immune system get added on to the earlier, innate system? Why are there quite so many different types of cells, which all seem to be sending signals to each other, often with apparently contradictory effects? In short, as every medical student grappling with the details asks, why is our immune system so complicated?

The need to live in close proximity to a vast assembly of microorganisms is the answer, say a growing band of researchers. They want to reframe our account of the immune system to take account of a wealth of new findings related to how we talk to our microbiota, and how they talk back.

Diplomatic relations

If you need a defence system to combat pathogens, the innate immune system works pretty well by itself. Turning that fact into a question is the key to the emerging view of the more complicated adaptive immune system of vertebrates. Why did the much more complex immune apparatus we benefit from make its appearance? A new answer was first presented in a brief essay in *Nature* in 2007 by someone who is not an immunologist but who thinks about symbiosis – the long-time student of the squid bacterial system discussed in the last chapter, Margaret McFall-Ngai.[9]

She was dissatisfied with the rather ad hoc suggestions that had already been put forward. Vertebrates, with their adaptive immune systems (they first appeared in fish), often grow large, live long lives, and may nurture a single offspring. All these traits, some said, called for more effective defences against other organisms and could account for the acquisition of the expensive and complex adaptive immune system with its capacious cellular memory. But, McFall-Ngai pointed out, some invertebrates also lead long lives, grow large, and produce only one offspring each year, and they manage well enough without this new immune clobber.

She suggested that the explanation for its appearance lay

elsewhere. If you survey symbioses across all the creatures that we now know live with communities of microbes, there is a clear pattern. Invertebrates typically interact fruitfully with just a few species of microbes, and often only one. In quite a few cases (more than one in ten insect species) it lives inside the host's cells. That intracellular niche makes microbes invisible to the innate immune system. Other microbes that live in invertebrate guts, for example, are normally there because of random encounters, and are just passing through. Only in the vertebrates do we tend to see so many microbes, working in mutually beneficial collections of species we can call consortia. Only there can we demonstrate they have co-evolved with their hosts and achieved a stable tenancy.

Her hypothesis then followed: 'The evolution of the vertebrate immune system is likely to be strongly affected by the need to maintain a substantial resident microbiota,' she wrote. The suggestion was not widely taken up at first, though. In fact, she says, her essay was greeted indignantly by immunologists. But she now observes with some satisfaction that the idea has been developed by a number of those in the field. She can see why there was resistance to her suggestion. She was not herself an immunologist. And, she says, 'cell biologists have typically not focussed on evolutionary biology or animal diversity'. However, 'They had no satisfactory explanation for the evolution of such a Rube Goldberg sort of system' either.* You can see what she means in another ingenious explanation for the appearance of the early adaptive immune system

---

* UK readers, for Rube Goldberg please read Heath Robinson, the cartoonist designer of wondrously complicated machines to do everyday tasks.

as cartilaginous fish developed jaws. This meant they could chew up harder foods, some immunologists speculated, which tended to damage the gut wall and increase the risk of infections. Faced with this kind of ad hoc reasoning, McFall-Ngai's suggestion looks all the more attractive. As she maintains, 'the management of the consortia makes way more sense than any other theory'.

In this developing new view, the whole point of the elaborate system of antibody recognition and the network of regulatory cells and chemical signals is to allow more discretion in deciding what *not* to attack. The innate immune system alone works for an organism whose position is 'just say no', says McFall-Ngai. But one that is hosting a large collection of other organisms, ones that are in some sense still 'non-self', needs a much subtler approach.

Others who endorse her view include Sarkis Mazmanian at CalTech, who has dug deep into the complexities of immune interaction in germ-free mice repopulated with designer microbiota. He and his colleague Yun Lee argue that the microbes that dwell inside all our lives were a more powerful influence on evolution of the adaptive immune system than fleeting encounters with the disease-causing organisms we used to think were the main players. They add another speculative twist: 'Symbiotic microbes may have influenced features of adaptive immune system evolution and function more profoundly than pathogens, possibly *to protect both host and microbiota* from invading infections'[10] (my emphasis there).

The adaptive immune system, in this view, co-evolved with the more generously populated microbiome. It maintains a delicate, dynamic balance. The microbes are allowed to stay because they bring advantages to their host, but they still have to be kept in check. They must be in the right place, and not be allowed to grow elsewhere. And potentially harmful ones still have to be eliminated. On the other hand, while the microbes need to be monitored, the immune system has to be reined in, and prevented from overreacting to the presence of all these different organisms. Failure on that side of the equation means permanent, unnecessary inflammation, and very likely disease.

Beyond dualism

How best to sum up this new way of looking at the immune system? Stephen Hedrick of the University of California at San Diego described McFall-Ngai's proposal as 'The notion that the immune system plays the bouncer, raising the velvet rope for beneficial bacteria and giving attitude to their less desirable brethren'. A number of other immunologists have offered more detailed versions of the theory now, each tending to throw in their own preferred metaphors to fill the gap left by avoiding talk of all-out war against invaders. The ideas are still developing, but one helpful way of looking at the big picture comes from Pasteur Institute researcher Gérard Eberl. He cuts through the details with a simple three-stage account of the development of ideas about how the immune system works. And he argues that our more detailed understanding of microbe–host interactions means we have to abandon simple ideas about what defines a pathogen, as well as a one-dimensional view of immunity.

Eberl points out how the original idea that adaptive immunity is there to discriminate between self and non-self had to be modified. There are too many examples of foreign agents failing to provoke a response. It is a bit more helpful, he reckons, to see the immune system reacting primarily to danger signals, as proposed back in the 1990s. It is still educated to be tolerant to self, but the self is redefined to include, for example, microbes that confer benefits on the host.

But this does not go far enough, he argues. It is still basically a dualist theory. Everything revolves around judgments of good and evil. 'Good includes normal self and mutualistic microbes, whereas evil includes altered self such as dead cells releasing danger signals and pathogenic microbes that alter the antigenic landscape of normal self'.[11]

There is a big problem fitting all the detailed results about the minutiae of cellular response into this framework. Maintaining the dualist view means you have to invent two classes of inflammation. There must be a normal, physiological or homeostatic level of inflammation that helps keep everything stable by, for instance, helping make sure the intestinal microbiota stay where they should and do not stray. Injured tissues or the presence of pathogens trigger a sharper response and lead to 'full-blown' inflammation, as the familiar short-term reaction to a problem.

That just does not work as a description of what we actually see, Eberl insists. There is no duality, but *continuity*. Like a literary critic or a cultural theorist, he depicts a cellular world in which everything depends on context. In the continuum model of the immune system, Eberl writes, 'microbes navigate between shades of good and evil'. The precise shade is determined by interaction with the host, and can change with host, tissue, and time. The immune system shapes the microbial environment to allow the organism to live with the microbes. It is not a fight between good and evil, 'it is rather an equilibrium between microbes and host that generates a superorganism'.

It is a superorganism that lives in dynamic equilibrium. An

organism, like us, that lives with a continually renewed population of microbes, needs to maintain a balance between hospi-
tality and hostility. That balance can be altered by influences
from the environment – in the shape of chemistry, food and
microbes entering the system. Or it can be affected by changes
in the host, caused by mutation, injury, or other types of stress.
Organisms faced with this problem needed to evolve a system
that (usually) responds in a way that maintains equilibrium in
the system in the face of these changes. 'The immune system
perfectly matches this function,' Eberl suggests.

He emphasises the dynamic aspect of all this with a
mechanical metaphor. 'In the superorganism, the immune
system is never at rest. It is like a spring: the more microbes
colonize the host niches or behave like pathogens, the stronger
they pull the spring of immunity, and the stronger the spring
of immunity pushes the microbes back. In germ-free animals,
the immune spring is close to rest, but in animals grown in a
normal microbial world, the immune spring is always under
tension, the tension required to maintain homeostasis.'

Here's one series of experiments yielding results that fill out
this picture, showing how the immune system helps fashion
the niches that microbes occupy. If you dose mice, or humans,
with antibiotics that wipe out most of the rich microbial life
of the intestines, they are in grave danger of getting infected
by an antibiotic-resistant strain of *Enterococcus*, a common resident of the gut that occasionally turns nasty. Normally, other
organisms living in the gut send signals that cause epithelial
cells lining the intestine to pump out antimicrobial peptides,
which keep down the population of *Enterococcus*. Experimentally,
production of the in-house antimicrobials can be restored in
mice simply by administering LPS, a molecule found on the

cell surface of the missing bacteria.[12] That is recognised by a receptor on host cells that form part of the innate immune system and starts up manufacture of the antimicrobial peptides. As Eberl puts it, the symbiotic microbiota 'pulls the string' of immunity. Whether operated by string or spring, the point is that the antimicrobials help fashion an intestinal niche that allows symbiotic bacteria to flourish but is toxic for potential pathogens such as antibiotic-resistant *Enterococcus*.

Developing a sense of balance

You can find support for this new picture of the immune system in a wealth of other experiments that illuminate how the microbiota and our own cells talk to each other. It turns out that there is more to the relationship between microbial communities and the immune system than co-evolution. They also mature together in each new individual, each influencing the other. The important thing for understanding our own health is that there are critical periods when some of our favourite microbes are necessary to trigger normal development of the immune system. Before looking at how this works, let's gather a little more detail about the main vessel for microbes that live with mammals: the gut.

As I said in Chapter 5, the gut has to manage two conflicting requirements. It absorbs nutrients and other small molecules across its expansive epithelial surface. But although it wants to maintain trillions of microbes in the colon to help feed our cells' appetite for those molecules, it needs to keep them out of the rest of the body.

That is achieved by a combination of straightforward barriers and the more subtle operation of the immune system. The barriers themselves are reasonably effective. The epithelial cell layer has those tight junctions. And the inner surface of the gut is covered with mucus, in a layer that gets thicker as you move down into the colon. Specialised cells in the epithelium secrete mucin proteins that have complex carbohydrate molecules added on and form a watery layer that sticks to the epithelial surface. In fact, there are two layers, with different, complex structures that still are not fully understood. The innermost layer, at least, is normally microbe-free.

Along with the cells that manufacture mucus, the gut has immune cells that are the other vital contributors to the environment in which the gut microbiome lives. As I said, the gut is the largest immune organ, accounting for perhaps 70 per cent of all the immune cells in the body. They are in and beneath the epithelium in a large complex of cells helpfully labelled gut-associated lymphoid tissue (GALT). There is plenty of cellular diversity involved, but let's skip that. What matters is that there is a continual traffic of signals in both directions between these immune cells and the microbes outside the epithelium. And the majority of this traffic is between immune cells and non-pathogenic bacteria.

Some of the messages passing to and fro can be isolated by experimenting on germ-free mice, and varying the microbes they are allowed to encounter, the genes their own cells express, or both. The results still depend on a complete living creature, but careful design sheds light on detailed links in the cellular communication networks in the gut.

The first big thing researchers established was that if they stay germ-free, mice grow up with a stunted immune system in the gut. All the bits and pieces of GALT – which include various lymph nodes and specialised patches of immune cells – remain underdeveloped, and the mice make many fewer cytokine cell-signalling molecules, grow fewer lymphocytes than usual, and pump out less immunoglobulin. It is as if the whole apparatus for talking to colonising microbes stays quiet because the immature gut senses that there is no one to talk to.

Inoculating an infant mouse with normal mouse micro- biota can restore immune function. More surprisingly, using human gut microbes – or even rat microbes – does not work. The effect is species-specific. In fact in some cases, a single species of microbe will do the trick. The overall effect is not a simple one-time trigger that leads to normal development, though. There is always some immune activity, and control sig- nals continually propagate through the system. Until the gut has been colonised, the regulatory signals are mostly ones that tone down immune responses, which encourages development of the microbiota without inflaming the epithelium. Among other things, the colonising bacteria then promote coordinated production of plentiful mucus, antibacterial peptides, immuno- globulin and immune cells in a collective that has been called a 'mucosal firewall'. The microbes secure peaceful co-existence by bringing on the conditions that ensure their own containment.

Once the gut is full of microbes, they are continually sam- pled by specialised body cells. A few will get across the epithe- lial wall, but most are scanned by the specialised white cells we have met before known as dendritic cells. These are mainly found in dome-shaped structures known as Peyer's Patches, depots for immune cells that are scattered over the gut epi- thelium. Dendritic cells are active sentinels. They can poke a thin arm-like extension across the epithelium and the mucus layer (thinner over the patch), and grab a piece of another cell. They then bring their booty – an antigen, or sometimes a whole bacterial cell – back inside the patch and present it for inspection to still-maturing lymphocytes. Some of those are then prompted to develop into regulatory T cells, which make a big contribution to setting the level of immune activity.

These intracellular mechanisms have also lent support to recent ideas about the point of that curious pouch that sticks out from our large intestine – the appendix. Ours is relatively small compared with other mammals and was long regarded as an evolutionary remnant. Surgeons often whipped it out, not just when it got infected in acute appendicitis, but as a precaution during bowel surgery undertaken for other reasons, to prevent future problems.

However, as Bill Parker of Duke University has argued, it is usually filled with normal gut bacteria and may be a reservoir of helpful species that comes into play when the rest of the gut gets depopulated by illness. It could, he suggests, 'reboot' the gut after dysentery.[13]

That could be true. But there may be another function, too. The appendix is also well supplied with immune cells, and is thus one of the key locations where the superorganism establishes which bacteria are welcome members of the gut community.[14]

The tangle of overlapping signalling networks in the gut – and other tissues where the immune system comes face to face with the microbiome, like the skin or the mouth – is being unravelled, slowly. Many cells must receive a variety of prompts to an array of receptors at any one time, and somehow integrate the information they represent. So an individual immune cell will be more or less active depending on the state of a collection of different receptors on its surface. It is a bit like the way a neuron in the brain 'decides' to fire depending on the balance of input from activating and inhibiting synapses from the other neurons in the same net, except that the primary signals are chemical rather than electrical.

Some of the interactions being uncovered are quite specific. One important brand of T lymphocytes, T-helper17 cells, are induced to develop in the intestines (of mice, anyway) by organisms known as segmented filamentous bacteria, which can cling on to mucosal surfaces. We cannot be sure the effect is completely specific. Some types of these bacteria certainly do other things that affect the immune system. Other bacteria could have the same effect on this kind of T cell, and hence on the production of the important pro-inflammatory cytokine it produces. But at the moment that looks unlikely. Laborious testing of complex microbial mixtures in germ-free mice shows that the effect is found only if these particular bacteria are present.[15]

In healthy folks, the overall result of many such interactions, once a stable microbiome has been established, is a kind of Goldilocks effect. The immune system remains active and on the alert, but not so alert as to be hypersensitive – it is just alert enough. There is neither too little, nor too much inflammation, but just enough. This balanced state is described by immunologists as homeostasis, adopting the term physiologists use to describe self-regulating systems in tissues and cells. Somehow, the immune system sorts out the welter of signals indicating the presence of food molecules, helpful bacteria, pathogens, and our own body cells, and reacts appropriately. The pattern of responses is laid down partly by our genes, partly by early interaction with pioneer colonisers. The full story here is not known, but a best guess is that these interactions have lasting effects partly via enzymes that alter key immune cells epigenetically, that is, by adding or removing chemical groups to their DNA that allow particular genes to be turned on or off.

A failure of education

The new view of the immune system lends itself to cosy conversational metaphors – researchers now write about negotiation, diplomacy and cooperation. But the old-style war metaphor contains one large truth. Our immune systems do incorporate some formidable weapons. Look at what a neutrophil can do, for example. These versatile cells circulate constantly in blood and lymph, and migrate to sites where the immunological alarm has been sounded. Once there they can engulf pathogenic bacteria, and kill them. They also release antimicrobial peptides, harming cells they have not managed to bring inside them. Finally, a neutrophil can expel an assembly of nasty agents bound up in a web of DNA and protein known as a neutrophil extracellular trap, which also immobilises and then kills bacteria. It is like having a guard dog that can bite, spit poison or, if enraged, fashion a weapon of last resort from its own insides and hurl them at intruders.

These armed and dangerous cells have to stand ready for use – because we live in an environment awash with microbes of all kinds – but not be allowed to get too excited in their daily round. We need our guard dogs to be vigilant, but on a very tight leash.

That is not an easy arrangement to contrive in a world of cells and molecules in motion, and it depends on careful education as well as native endowments. The sometimes drastic alterations in the modern microbiome, some fear, are compromising that education. Bypassing the vaginal bacteria at birth. Treating infants with antibiotics. Being frugal with fibre but lavish with fat and sugar. Even eliminating parasites like roundworms. All change the conditions in which the microbiome and the immune system learn the lines that script their performance.

And as those conditions have changed, so has our health. Overall, the story is one of improvement, from the germ theory on. But it is undeniable that we still get ill, and the pattern of illness has changed. And when we look at the long list of conditions that have been linked with the modern shifts in the microbiome, almost all are also linked with a derangement of the immune system. Some are autoimmune diseases, in which immune cells whose programming has been messed up destroy cells in our own tissues. Some are affected by what has become something of a catch-all explanation for ill-health, chronic inflammation in one tissue or other. In either case, suspicion arises that the conversation between the microbiota and the immune system has taken a nasty turn. In the words of one recent review of all these interactions, 'alteration of the composition and function of the microbiota ... has transformed our microbial allies into potential liabilities'.[16]

How much of a liability are they? Time to look at some of the possible medical consequences.

8 | There goes the neighbourhood

Nausea grips you. There is a greasy sheen of sweat on your brow, but you feel shivery. There is an ominous grumbling in your guts, and you sense uneasily that you know what is coming. If you are lucky, you make it to the loo in time to vomit up the contents of your stomach through your mouth and nose. You have been sick. Really, really sick.

The Old English (before that, Old Norse) word *sick*, for illness in general, became used for throwing up some time in the 17th century. There is something about vomiting's sudden grip on your whole body that makes it one of the purest experiences of being ill. There are plenty of ways of being unwell, but being 'sick to your stomach' is one of the hardest to ignore.

In the 21st century, we are finding many more correlates of sickness in our guts. When we are ill, science looks for answers in our organs, tissues and cells. As we now appreciate that they are all part of a superorganism, researchers are exploring how our microbiome may be involved in illnesses of all kinds. In fact, a majority of our ailments have been linked to our microbiota in one way or another. Sometimes the connection is direct, and we are beginning to understand the chains of cause and effect in detail. More often it is indirect, and it is hard to tell whether changes in the microbiota are the cause of an illness, a consequence, or just happen at the same time. This chapter looks at what we know so far about the microbiome in health and disease, and where the research is heading. The place to start is obvious. The vast majority of our microbes live in the colon. And there are some pretty nasty diseases of the colon. Could they possibly be connected?

Fire down below

As we try to understand how the microbiome gets involved in disease, we must stay alert to the possibility that immune reactions, notably inflammation, are one aspect of this that ties many conditions together.

With the inflammatory bowel diseases (IBDs), this connection is a given. They involve inflammation of portions of the gut. But why does it happen and what we can do about it?

A lot of people want the answers. The main serious conditions under this heading – Crohn's disease and ulcerative colitis – affect a quarter of a million people in the UK, a million and a half or more in the USA.* The severe bowel conditions can cause extreme discomfort, and worse. The comedian Stewart Lee expressed himself quite strongly about what the worst form of ulcerative colitis, diverticulitis, felt like, recalling, 'my bleeding arse, my orange piss, my contorted stomach, watching my blood bubble around the needle in the drip, feeling the cool saline fluid in my veins, suddenly

---

* IBD is distinct from irritable bowel syndrome (IBS), another common, chronic source of intestinal discomfort that is not linked with inflammation. There are lots of studies of the microbiota in IBS, too, but without any very clear picture emerging so far.

losing and gaining weight, feeling like a tube of screaming meat whose only purpose was to process the muck I ate and crap it out the other end.'[1]

These morale-sapping disorders are also a good place to begin looking at microbiome effects because some things about them are well understood. We know quite a bit about the involvement of particular types of immune cells, and signalling molecules, in unleashing the attacks that lead to an inflamed colon. And the mechanisms at work here look like a promising model for how inflammation is controlled, or gets out of hand, as part of the maintenance or breakdown of homeostasis in the larger system that is the colon and its contents.

You can see the power T lymphocytes have to damp down the immune system by what happens when they are not there. One kind of regulatory T cell produces the cell-signalling molecule or cytokine called interleukin 10 (IL-10), which carries a general message to other immune cells to keep calm and carry on. Experiments on mice in the 1990s confirmed that modifying or blocking production of these cells produced inflammation in their intestines that looked like human bowel disease. This was a genuine breakthrough, and furnished a model for the condition that has helped unravel some of the molecular interactions involved in the control system. At this level, we know a lot of details about which kinds of immune cells send and receive signals that regulate inflammation in the gut by producing either pro- or anti-inflammatory cytokines, and how the system gets out of kilter. If Crohn's or colitis are severe, there are now powerful treatments – very effective for some – that block cytokines to reduce inflammation. They have to be trialled carefully because messing with cytokines often has effects that are miles off target, but finely tuned inhibitors of production of inflammatory cytokines, for example, can do the trick.

There is also a long-standing speculation that changes in our complement of colonisers have promoted an increase in Crohn's disease. This originally focused on parasitic worms, which still afflict much of humanity and may well have always done so. As they can have lots of debilitating effects, countries lucky enough to have modern sanitation and health systems have eliminated them. This change has coincided with a leap in diagnosis of Crohn's disease, leading some to put two and two together. Could worms, or their eggs, help train the immune system not to attack the gut epithelium? Some Crohn's patients have been desperate enough to test this theory by infesting themselves in hope of relieving their symptoms.[2]

But now we can also ask, what is the role in all this of the bacteria that are a hair's breadth away from the delicate tissues of the gut epithelium and its immune networks? There is plenty of evidence that suggests they are deeply involved. One impressive line of work was reported from Sarkis Mazmanian's laboratory at the California Institute of Technology in 2010. It involved another striking experiment using a cell-surface molecule like LPS (mentioned in the last chapter as capable of inducing production of antimicrobial peptides that help ward off *Enterococcus* infection in mice).

Mazmanian's group were also the researchers who showed some years earlier that a single much-studied bacterial species, *Bacteroides fragilis,* can prod germ-free mice into developing a normal gut immune system. Now they demonstrated that the same bug acting alone can induce the development of specific regulatory T cells. In fact, a single antigen from the bacterial wall, polysaccharide A (PSA) can do the same. *B. fragilis* is doing this not because it cares whether the bowel gets inflamed or not, but because it suppresses an immune reaction to its own colonisation, allowing it to nestle up against the intestinal mucosa. But in a normal, mixed microbiota this would be one of the interactions that makes for balance in the immune system. The obvious next test had an impressive result. Feed PSA to mice with inflammatory bowel disease and their symptoms are reduced.[3]

Earlier, complementary experiments had shown that a different bacterium, *H. hepaticus*, promotes *inflammatory* cytokines so effectively that it can be used to produce another good mouse model of IBD. But *B. fragilis* opposes the morbid influence of *H. hepaticus*, with the anti-inflammatory agent it evokes winning the battle of the cytokines.

These experiments begin with germ-free mice. In a normal mouse, as in a human, the bacterial species used are only two among hundreds or thousands. It is unlikely that they are the only species trading in these kinds of signals, so suggestive work like this has to be interpreted in the context of studies of the total microbiome.

The overall microbial population does change in IBD – often said to be one of the conditions that is associated with 'dysbiosis'. That is an elusive term. Sometimes it denotes specific changes in microbial populations, but it is often used more loosely. Sometimes it just seems to mean a microbial mix that looks a bit different from normal.

As you will be expecting by now, trying to get beyond experiments with germ-free mice and get some clear results from human studies plunges you deep into the complexities of shifting bacterial populations, immune responses from human tissue, diet, and other factors like stress.

We would like to find 'biomarkers' – unique indicators of the disease, for diagnosis or prediction – or even particular microbiome components that cause a problem and can be tackled, by elimination or substitution, to relieve symptoms. What we get instead, so far, is a mass of data which, at best, can be resolved to produce 'signatures' of the condition. What this tells you is basically 'if you have Crohn's disease, this is what your microbiome looks like'. Getting from there to a clear idea about what causes what, and what might be done about it, is not straightforward.

A nice example of the early efforts to move from studying the microbiome of healthy people to understanding its role in this disease was a small twin study published by a group led by Alison Erickson of the Oak Ridge Laboratory in Tennessee in 2012.[4]

They already knew that there are differences between the gut microbiome in people with Crohn's and those without. Twins show this clearly – they tend to have similar microbiomes, in terms of species distribution, even when they have lived apart for decades, but there are still marked differences if one twin has developed Crohn's disease and the other remained healthy.

Erickson's lab chose six pairs of twins for closer study. To cover all bases they included a healthy twin pair, one in which both twins had Crohn's disease in the colon, two pairs in which both had the disease in the ileum (the last stretch of the small intestine) and two pairs in which one had the disease and one did not.

They collected stool samples from each, and compared results from a three-pronged analysis. As well as identifying which microbial species were present, they wanted to know what they were capable of, metabolically speaking, and whether they were doing it. The metabolic investigation meant going further than the basic 16S rRNA readings and probing the DNA sequences of the mixed-up genomes of all the different microbes – the full metagenomics analysis. As we saw in Chapter 1, then they would have a good idea what proteins, especially enzyme proteins, each microbiome could make, and hence what chemical reactions the bacteria could promote, working together.

Knowing that is interesting, but it is better still to know which of those proteins are actually being made. That was the third part of their plan. That information, not surprisingly, comes not from DNA analysis, but from another of the many new varieties of '-omics' – proteomics. That means identifying all the different proteins being made in a particular cell, or collection of cells. In this case it was actually done by applying the long-established technique of mass spectroscopy. As the name suggests, this separates molecular mixtures by size (or mass). If you apply it wholesale to an extract of bacterial gloop you can compare the results with reference data in (yet another!) electronic archive and get a snapshot of the proteins actually being made. These protein snapshots are recognisably different for healthy subjects and those with Crohn's, and the differences more marked in those with ileal disease.

These were shapes in the mist, rather than clear results, though. There is a lot of what you might call molecular noise in the data. These few subjects, for example, yielded more than 1,200 proteins that were unique to those with healthy bowels, another 700 unique to those with ileal Crohn's disease and 145 more seen only in those with colonic Crohn's. That is a lot of difference in gene expression from a change in the state of the microbiome.

Among all these proteins, many had been seen before, but around 30 per cent have no known function. So for all the precision of the analysis, the main thing it tells us is that people with ileal Crohn's disease have a gut microbiome making a smaller range of proteins. Frustratingly, where function is known, this applies across lots of different types of proteins. Their list covers enzymes for carbohydrate transport and metabolism, energy production and conversion, handling amino acids and lipids and most other common cellular functions. Basically, the results show that if you've got Crohn's disease in your ileum, things are pretty messed up. But if you have the disease, alas, its effects have probably told you that already.

One day, a better understanding of the information represented by these 'dozens of species, thousands of metabolites and hundreds of proteins that vary in relative amounts' may lead to more carefully tailored individual treatments. Before then, they ought to lead to better diagnosis, which would be a boon for many patients whose early symptoms are hard to read. That is because early symptoms of Crohn's, which include stomach ache and diarrhoea, often lead to doctors prescribing antibiotics. Bad idea. A larger-scale 2014 US study looked at patients newly diagnosed with Crohn's disease, and confirmed that they have less diverse populations of gut microbes, and an increase in the proportion of organisms that seem to provoke inflammation.[5] Children who had moved on to a diagnosis of Crohn's had a more disturbed microbiome if they had been taking antibiotics beforehand than those who had soldiered on without. The first-line treatment might have gone on to make their symptoms worse, in other words, although that did not show up in the timescale covered by this study. But it is another addition to the evidence that antibiotics should be used with caution.

So the search for generally applicable ideas comes back to inflammation, and the fact that it can have multiple causes, including a response to some gut bacteria. Do messed-up microbiota 'cause' inflammatory bowel disease? Probably not usually, or not alone. What does look very likely from further mouse experiments is that a skewed immune response tends to favour bacteria that can grow in an inflamed colon, and which themselves promote more inflammation. Mice with immune systems that are altered in several different ways that make them prone to ulcerative colitis develop a characteristic microbiome. Transfer that microbiome into normal mice, and they get the disease too, although not so seriously.

The microbial mix in question probably promotes inflammatory disease from both directions, as it were. It is rich in species that exacerbate immune activity. At the same time, it has few of the bacteria that digest complex carbohydrates, whose short-chain fatty acid products have a calming effect on immune cells, as I described in the vignette about butyrate in Chapter 5 (*see page 105*).

This remarkable development of a gut microbiome that makes IBD contagious is the kind of experiment you cannot do on people. But it suggests that our own gut microbiome is at least likely to be involved in maintenance of IBD once it gets under way, making it harder to shake.

Mass killer, small culprit?

Moving away from the bowel, the gut microbiome is involved with quite a few of the risk factors for the biggest cause of death in Western societies, cardiovascular disease – that is, heart attacks and strokes. Most of the links are subtle, and are tied up with conditions that have complex origins like obesity, diabetes, and chronic inflammation. One that looks more direct is a recent discovery, and highlights the worrying fact that not all the bacteria in the gut are doing us chemical favours.

Some gut bacteria, going about their own metabolic business, turn a set of dietary nutrients into a compound called TMAO (trimethylamine-N-Oxide). Stanley Hazen of the Cleveland Clinic and colleagues have shown that TMAO levels in the blood are a good indicator of heart attack and stroke risk, and that TMAO promotes clogging of blood vessels in mice.[6]

Compounds in food that generate TMAO through bacterial action include lecithin (also known as phosphatidylcholine), found in eggs and red meat, and carnitine, also featuring in meat and dairy foods. Hazen's studies confirm that bacteria are responsible by showing that blood TMAO increases after eating eggs, but not if the diner has had a course of antibiotics first. Vegetarians produce less TMAO even when they eat carnitine, suggesting that meat encourages high populations of the bacteria that do the damage.

Perhaps this will explain why some people succumb to stroke or heart attack who do not score high on other risk factors such as blood cholesterol levels. A gut microbiota profile may one day be used to advise people whether to avoid foods rich in TMAO precursors – though like cholesterol they are indispensible in modest amounts. Probiotics could also moderate TMAO production, but which ones might work remains to be investigated.

Meanwhile, it is an unwelcome addition to the ways our microbiome can affect the risk of a major medical problem. But there are plenty more of them.

My *Firmicutes* made me fat ... or did they?

What makes people overweight? We know something surely is. One of the less happy signs of the times is that for the first time there are more obese than underfed people on the planet. The figures are alarming. Public Health England says that the prevalence of obesity in England tripled – from around 7 per cent to 21 per cent – in 30 years from 1978 to 2008. In the USA, on the same definitions, obesity rose from 15 per cent to over 30 per cent of the population over the same period.

What is going on? We have easy access to more, and more fattening, foods and drinks than most people have had in the past (and new ingredients like high-fructose corn syrup). Fast food is cheap, tasty, and cleverly designed to push evolutionary buttons set for a time when salt, fat and sugar were in much shorter supply. This is what public health experts call an obesogenic environment – a word as unappealing as some of the consequences of what it is trying to explain.

Still, not everyone gets fat. So can those of us who stay lean congratulate ourselves on our willpower? What else might differentiate the hefty from the slender?

Genes might be one indicator. They do seem to shift the probabilities. Much research has found that there are gene variants that occur with different frequencies in lean and obese people. But the best genetic information still only allows a prediction of whether the owner is obese with 58 per cent accuracy – not a lot better than flipping a coin.

Everyone sat up and took notice when a different gene scan, of the gut microbiome, allowed much better predictions. Microbial species, too, differ between lean and obese people, and more markedly than their genes. If you scan the genes in the gut microbiome, you can increase the accuracy of the lean versus obese prediction to 90 per cent.

So your bugs make you fat? Well, once again the headline finding is the start of a more complicated story. Looking at the gut microbes of people who are already obese cannot show which came first: the weight gain or the change in the microbial mix.

Since this field only really got going a decade ago, there is not yet much in the way of long-term studies of people gaining or losing weight that tracked changes in their microbiomes to try to disentangle causes and effects. What we do have are a clutch of short-term studies in people, and a growing number of experiments with mice. There are also re-examinations of data from long-term studies of child development that help frame the whole discussion. Children who are dosed with antibiotics when young are more likely to grow up obese. This has been found in 'birth cohort' studies – in which babies are followed for decades and future outcomes correlated with detailed medical records – in Denmark, the UK and Canada. The starkest results come from the Canadians, who reported that infants who were prescribed antibiotics before their first birthday were nearly twice as likely to be overweight at age twelve as those spared the drugs (32 per cent versus 18 per cent). This does not tell us what might have changed in the microbiome to produce this outcome, though. And it poses a fresh puzzle. When the researchers corrected for a collection of other variables, including birth weight, breastfeeding, and whether the mother was overweight, the difference disappeared in girls, but was still visible in boys.[7]

Such a puzzle will be solved only by studies that get at the microbiome directly. Some significant things emerge from all these – both suggestive facts and results that modify some simple hypotheses about microbes and body mass.

The study that first got everyone's attention came from Jeffrey Gordon's lab at Washington University in the USA. To begin with, they worked with mice that had grown up fat because they did not make the hormone leptin, which helps regulate appetite. The microbes in their guts, on analysis, proved to be a different mix from those found in normal lab strains. Specifically, they had a lot more varieties of *Firmicutes* (F), a lot fewer *Bacteroidetes* species (B) and less microbial diversity overall.[8]

They went on to find the same population shift in people, the F/B ratio distinguishing between lean and obese humans, and shifting back toward the 'lean' setting when obese people managed to lose weight.

Can we confirm that this is what normally happens? Well, in other studies these basic findings seemed to hold up whether you investigate obese and lean adults, overweight pregnant women, children who develop obesity, overfed rats, or rats genetically engineered to be obese.

But. Keep reading the studies and you also find results that show no difference in the F/B ratio between the lean and obese and even a reversal of the effect. The picture blurs.

The broad shift in bacterial population did not make obvious sense in terms of mechanism, either. Food has to be digested. We know bacteria in the colon help digest otherwise unavailable foodstuffs. So the most obvious early hypothesis was that having the altered mix of species in the gut meant obese people were recovering more calories from their food, specifically from complex carbohydrates. This aligns with the fact that germ-free mice have to eat more to achieve the same growth as their normal counterparts.

We cannot (easily, ethically) substitute one lot of bacteria for another in people, but mice don't mind. And this idea did not pass experimental tests. In any case, it did not seem very logical. People seem to become fat not from chewing their way through lots of complex carbohydrate, which probably featured more in the diets of our ancestors, but from eating lots of fat and refined sugar in the foods we now find hard to resist. But these, we think, are digested in the stomach and intestines, largely without microbial assistance. However, take germ-free mice and rear them on a similar high-fat, high-sugar diet, and they do not gain weight like their cousins with a full complement of gut bacteria. So it looks as though there is something else our microbes are doing beyond simple breakdown of large molecules that can help to make us fat. But what?

One strategy to regain clarity is to focus on smaller subdivisions of bacteria. *Firmicutes* are a whole phylum – the large evolutionary category. Perhaps there is a class, or even a species, within the phylum that is particularly important? Tracking shifts when people lost weight might offer some clues here.

Obese people who manage to slim down see their gut microbiome change. That happens in weight loss by a range of methods, including diets – sometimes pretty drastic ones – and even gastric bypass surgery, which rejigs the actual anatomy of the digestive system to reduce nutrient absorption.

The changes are not always the same. Some individual case results have been spectacular. A morbidly obese man who shed 51 of his 175 kilos in four months in one widely publicised study at Shanghai Jiao Tong University in China saw a radical microbiome shift from 35 per cent *Enterobacter* before to negligible amounts after weight loss.

The same *Enterobacter* fed to germ-free mice also caused

them to gain more weight on a high-fat diet than control mice, although not on regular mouse food. Here we encounter the perhaps unexpected influence of the immune system in the regulation of body mass. This particular bacterium produces one of the lipopolysaccharide cell wall compounds that is recognised by the innate immune system. Among many other effects it can induce insulin resistance and obesity when injected into mice, in a reaction bound up with the inflammatory response.

So this single patient may have gained extra weight because he harboured this bacterium – and lost it partly through eliminating the microbe, as well as thanks to a radically controlled four-month diet of 'whole grains, traditional Chinese medicinal foods and prebiotics' that helped see the bacterium on its way.

However, while the same diet for nine weeks in a human trial with more than 100 people did produce some weight loss, and bacterial changes, the first subject proved to be an extreme case (as was the diet, by Western standards, incidentally).[9]

Intriguing though this is, it is one study among many. The microbiome is complex and there is unlikely to be a single species behind obesity. Indeed, there probably is not a single cause of obesity, either in general or even, usually, in a single person. Any effort to map all the links of cause and effect quickly gets pretty tangled.

Martin Blaser and Laura Cox cut through this by simplifying things down to four main influences.[10] Obesity is promoted, they say, by a high-calorie diet, genes, shifts in physiology, and the composition of the microbiome. Complications arise because any of the first three can act independently of microbial effects, *and* they can also be modified – for better or worse – by the microbiota. When the development of normal gut microbiota is disrupted, by delivering a baby by caesarean or by dosing infants with antibiotics (Blaser's special interest), they may be more inclined to obesity later on.

Those varied causes lead to the main result via various effects. In this scheme, the authors simply label the microbe-independent ones 'multiple host mechanisms'. The others, that are influenced by the microbiome, include changes in energy harvesting, metabolic signalling and, invoking the immune system, inflammation.

That all makes for a (reasonably) neat diagram in their paper. But the three main classes of effects are each themselves varied, and are not necessarily independent, so it still overlooks further complexity.

However, it is worth noting one scenario in which the causality loops back the other way from the possibilities they summarise. It concerns one of the main reasons excess body mass worries your doctor – the complex of changes known as metabolic syndrome. This is often linked to obesity, especially a roly-poly middle, and refers to a combination of high blood levels of fatty triglycerides and sugars, low levels of high-density lipoprotein (popularly known as 'good' cholesterol), and high blood pressure. Together these are a perfect storm of risk factors for type 2 diabetes, heart disease and strokes. The whole syndrome is also linked once again with inflammation, especially associated with fat-storing cells as a source of the activating varieties of cytokines. In that case, low-grade inflammation might be a consequence of metabolic syndrome.

Alternatively, inflammation stemming from other causes – including shifts in the gut microbiota – could be a prelude to metabolic syndrome, interfering with the action of insulin, for example. That fits with some studies showing microbial population shifts in patients with fully developed type 2 diabetes, the increasingly common kind that is due to cells failing to respond to insulin. The metabolic changes that are a feature of diabetes are large, though, so it would be surprising if this was not reflected in microbial signatures.

Type 2 diabetes is distinct from type 1, which results from depletion of the cells that make insulin in the pancreas. That happens when they are attacked by the immune system. And like a number of other autoimmune diseases, that has also been linked with changes in the microbiome by some researchers. Here the new view of the immune system outlined in the previous chapter may dovetail with much other work trying to understand conditions whose origins are often very hard to pin down.

Friendly fire

The new ways of thinking about the immune system prompted by consideration of its normally peaceful co-existence with trillions of microbes, discussed in the last chapter, automatically raise questions about what happens when things go wrong. Inflammation in the wrong place or at the wrong time – perhaps all the time – is one general adverse response to scrambled immune signalling. The other is when cells in our own organs and tissues become targets for activated immune weaponry. This kind of misdirection of a system designed to protect against threats to the body is the basis of a range of autoimmune diseases, of which type 1 diabetes is just one.

Autoimmunity is a very general idea, with hugely varied possible outcomes. In fact, the category of autoimmune disorders expanded in the 1980s and 1990s as more and more collections of symptoms were suspected of having an autoimmune component. The label is now applied to 80 or so different conditions, including Lupus, Addison's disease, Guillain-Barré syndrome, some anaemias, and more common conditions like rheumatoid arthritis. Some of the conditions are themselves very variable, and there is not always agreement about their classification as autoimmune syndromes. The least ambiguous ones are those where there are antibodies to our own cells that are specific to one cell type, the one that gets damaged.

Their origins lie deep in the heart of the adaptive immune system. The processing and selection of lymphocytes has gone awry in these diseases. Somehow, out of the vast array of possible antibodies that lymphocytes offer for inspection to see if they should become operational, one has been retained when it ought to have been discarded. Instead they circulate on the surface of immune cells. They may or may not then be activated when they meet their chosen antigen, which is often but not always a molecule found on the surface of cells of the vulnerable tissue. Some autoimmune conditions flare up and then die down again in unpredictable ways. They are not directly treatable or preventable. Doctors typically try a large collection of treatments to relieve symptoms or reduce the incidence of flare-ups.

We do not really understand how autoimmune conditions begin. So introducing a whole new set of variables like the population mix and metabolic and signalling networks of the microbiome has not yet produced sudden illumination. The microbiome is important for the development of the innate and the adaptive immune systems. We know their mutual regulation can be thrown out of balance, and lead to inflammatory responses where they will do no good. But the *specificity* of autoimmune reactions that target particular cell types in our own tissues does not fit easily into that story.

Many autoimmune diseases are linked to mutations in a complex of genes that code for cell-surface markers known as human leukocyte antigens. These mutations may increase susceptibility to autoimmune responses, perhaps by removing one level of safeguard. The full-blown condition may then be triggered by a secondary influence. This is where microbial effects look most likely to play a role.

Two hunches look like pretty safe bets before even looking at the literature on particular autoimmune conditions and the microbiome. There is a good chance that some connections will appear – autoimmune conditions seem to affect, and be affected by, almost everything. But there is little prospect that there will be any simple links that represent breakthroughs in understanding of these problems, or not yet anyway. Type 1 diabetes is a good case study here, partly because it is a little bit easier to get your head round than some of the other autoimmune conditions.

Happily, this kind of diabetes is manageable, although at the cost of taking over responsibility for your own homeostasis and committing to a lifetime of injections – so the burden is heavy. The fact that it appears when people are still young, and seems to be on the increase, means that a great deal of work still goes on to find better ways of dealing with it than monitoring your blood sugar levels and regular injections of insulin.

There are multiple theories for the origins of autoimmune diseases. They could all contain part of the truth, or simply be expressions of the theorist's favoured way of viewing the immune system. They seem to offer two ways in which the gut microbiota could plausibly affect the development of type 1 diabetes.

The first invokes the observation that gut microbes are

important to normal maturation of the immune system, and affect its regulation. If shifts in the microbial population make the immune system harder to control, that might be the way a child who is genetically at high risk for type 1 diabetes goes on to suffer destruction of the key insulin-producing cells. The research problem then would be to try to light up the pathways of molecular interaction that make the immune cells involved do what they do – in this case stop ignoring insulin-producing islet cells in the pancreas, and start attacking them.[11]

The other possible influence is more general, and involves an increase in 'leakiness' of the gut. As we have seen, the gut's barrier function depends mainly on a single thin layer of epithelial cells, with tight junctions between them. Sometimes, the junctions loosen up and allow more molecules to pass across the epithelial layer. This, too, can be part of the way the immune system works, permitting delivery of a crowd of antibodies, for example. But it also allows different molecules to go in the other direction, and dietary and other antigens to encounter immune cells. It is associated with bowel irritants like the painkiller ibuprofen, and with inflammation. There is evidence it is linked with inflammatory bowel disorders and diabetes but – does this sound familiar? – whether it is a possible cause, or is caused by the conditions, is uncertain. The same applies to connections between an increase in intestinal permeability and alterations in the gut microbiome – so what we are really doing here may just be increasing the number of variables that may or may not be connected causally, in either direction.

Even so, the circumstantial evidence carries some weight. Type 1 diabetes is one of the conditions that has increased in incidence in recent decades, and appears to be starting younger. Finland seems to hold the unsought-after record here, with a fivefold increase since 1950.

Could that be linked to alterations in the modern microbiome? Martin Blaser thinks this troubling possibility makes sense. But so far the evidence remains largely circumstantial. A detailed study in Germany in 2014 did find some differences in gut microbiome signatures between those children who had antibodies to the insulin-manufacturing islet cells in the pancreas circulating in their blood and those without. However, the difference was a subtle one, involving neither the diversity nor the number of bacteria in the gut, but what the researchers termed 'bacterial interaction networks'. The significance of this remains uncertain.

Reviews of the gut microbiota and type 1 diabetes also cover tantalisingly contradictory findings from animal models. The most recent review as I write reports on relatively few completed studies. A key paragraph relates that diabetes-prone rats are less likely to develop the disease if they are dosed with antibiotics. So, gut bacteria must make autoimmune destruction of islet cells more likely? But wait. The next study mentioned indicates that non-obese diabetic mice are *more* likely to develop full-blown diabetes in the *absence* of bacteria, so perhaps gut bacteria make autoimmune destruction less likely, as Blaser supposes. Or perhaps we just don't really know what's going on yet. That impression is reinforced by rat studies showing that probiotic strains of bacteria can either increase or reduce diabetic risk in animals, depending on the strain. The authors conclude, with as much scientific optimism as they can muster, 'Taken collectively, these results suggest that microbial exposures may play a role in the onset of the disease'. They do, however, hedge their bets in the conclusion, which begins by suggesting that 'unravelling the contribution of the microbiome in type 1 diabetes development may prove especially difficult'.

That does not mean there will not be a big effort to try. A substantial part of the review details larger human studies that are currently under way in Europe, North America and Australia. But the results will not be available for some years.

The current studies on diabetes certainly demonstrate that effects on the immune system are one route by which the influence of microbes in the gut *could* reach other parts of the body. That reinforces the interest in microbial links with other autoimmune diseases. Rheumatoid arthritis, for instance, is also under close investigation here.

It is another common autoimmune disease, in which immune cells suffer from the misapprehension that they need to eliminate cells in the joints. Here there is again reason to suspect a multi-step process. Extensive research has produced evidence for a wide range of genetic and environmental influences. Some patients come up with auto-antibodies well before there is any other sign of disease, so the search is on for secondary triggers.

At least one microbiome study offers a possibility. Dan Littman's group at New York University found that the gut microbiome of patients with recently diagnosed arthritis, as yet untreated, showed a big increase in the population of *Prevotella* species, led by one in particular, *P. copri*. This microbe was not found in such quantities in comparable groups who were healthy. But nor did they find it in arthritis patients who had been dealing with the condition for some time.

So far, it is another finding of difference that is intriguing, but needs further investigation. You might say the same about a finding that the gut microbes of patients with multiple sclerosis are enriched with *archaea* species. And one could go on to review studies of a clutch of other conditions that have hinted at links between changes in the microbiome and development of symptoms. I have focused on type 1 diabetes here because it seems to have produced the strongest research. But the general conclusion for now is that the microbiome, perhaps via the immune system, may be tied in with many of these conditions – but that it is too early really to tell.

Science does not of course always provide neat answers, though it is frustrating to be left with a collection of suspicions that may turn out to be merely suspicions. We can, perhaps, now savour uncertainty at a higher level. One idea that shows this progression clearly is a general suggestion that relates to the immune system, modern life, and the rise in suffering from a collection of conditions of the kind that trouble Martin Blaser – the hygiene hypothesis.

Down and dirty

Hay fever sounds like a timeless affliction, named, you might think, by people who spent summer in the fields bringing in the harvest. Wrong – it is distinctly modern. Sneezing and wheezing triggered by pollen is more likely to occur in the city than the country, and was only really noticed by doctors in the 19th century.

It has since become more common, one of another set of hard-to-fathom disorders that result from overreaction by our immune system – not autoimmune diseases, but allergies. Asthma, food allergies and eczema have all increased relentlessly in developed countries. Something linked with modern life seems to promote these annoying, debilitating, and some-times very serious conditions.*

In 1989, a Scottish researcher in London, Martin Strachan, published a paper analysing statistics from a very large study of child development that had begun in the UK back in 1958. He noted that coming from a small family, with few or no siblings, seemed to make it more likely for a child to develop allergies. Perhaps, he suggested, that was because they had fewer bouts of viral illness when they were very young – were less exposed to mumps, measles, or rubella. When he speculated that this was also due to higher standards of cleanliness, his explanation was dubbed the hygiene hypothesis.[12]

The idea was that evading these infections, mainly viral ones, was somehow skewing the immune system towards later overreaction to relatively innocuous stimuli. Before long, the idea was extended to autoimmune diseases.[13] The details were hazy, but it all seemed to fit some simple conceptions of the immune system. It is a temperamental creation, people seemed to suggest, sensitive to early rearing. Give it regular workouts, and it grows up happy and well-balanced. Deprive it of early experience, and it will emerge immature, sickly, and liable to lash out at the slightest provocation.

This little caricature fits the original hygiene hypothesis,

---

* Hay fever's emergence highlights the long-standing habit of ascribing a new or worsening medical problem to, well, whatever observers do not like about their immediate milieu. Mark Jackson relates how in the USA in the 1880s it was claimed to be a neurotic condition, generated 'by diverse pernicious features of modern life' including speed, noise, increases in the pace of business and discovery, and education of women. (Jackson, 2006). It is good to bear this in mind when reading explanations of more recent increases in illness.

but it also sits happily with the more recent evidence that the gut microbiota play an important role in overseeing development of the mature immune system. As changes in our microbiome have gone along with the reduction in infant viral infections, it then gets harder to separate out modern influences on the state of our immune responses. As researchers mulled this over, the hygiene hypothesis was supplemented by the 'old friends' hypothesis, due to Graham Rook of University College London. He suggested in 2003 that the rising tide of allergy and autoimmune disorders was due not to avoiding viral infections, but to lack of exposure to a more diverse range of microorganisms and parasites than early humans had to deal with.[14]

It is a small step from there to linking the hygiene hypothesis to the new findings now piling up about the gut microbiome. But that brings us back to the combined complexities of the immune system itself, and the microbial ecosystem that dwells in our colon. Yes, things have probably changed there. This has probably been registered in the development of your immune system, and mine. But the number of possible interactions is so astronomical, and the number of detailed experiments as yet so few in comparison, that we cannot be sure exactly what the effects might have been, or how we might adjust them if we needed to.

It seems likely, though, that we are looking at some of the consequences of these changes when we contemplate the rising incidence of a wide range of allergic, autoimmune and inflammatory conditions. The exact mechanisms remain largely speculative. At the moment the accumulating evidence from the microbiome does not allow us to make any very useful medical prescriptions. But it does allow us to speculate at a higher level.

Here is one rather delightful example, returning to Graham Rook. One way to help maintain a fully functioning immune system, he suggests, could be as simple as taking a walk in the park.

He offers two lines of thinking that converge on this conclusion. The first is his restatement of the hygiene hypothesis as a lack of old friends. When the human immune system evolved, he argues, it co-existed with organisms from our own microbiome, as transmitted from people in the family and supplemented by organisms from the environment that could join that community. There was a third influence, he assumes, from potential pathogens encountered by small, isolated groups of hunter-gatherers, which typically persisted as sub-clinical infections with few if any symptoms (parasitic helminth worms would be an example).

Unknown in our early evolutionary development were what he terms 'crowd infections' seen mainly in large populations in cities. The common childhood viruses fall into this category. Our modern cities still expose us to those, which activate the immune system, but we are helminth-free. Living in buildings made of modern materials also helps ensure that we have less exposure to the wider natural environment than at any time in the past. 'Until recently,' Rook points out, 'even our homes were constructed with timber, mud, animal hair, animal dung, thatch, and other natural products and were ventilated by outside air.'

Add another curious epidemiological fact. City folk who live close to green spaces, in leafy suburbs adorned with parks and gardens, are healthier on a host of measures. They live longer, are less likely to get depressed, and generally report higher well-being. Is this because green spaces encourage exercise? Could it be that they promote more social interaction? Perhaps we just find them appealing because they replicate some features of ancestral landscapes which we crave? None of these explanations is convincing, he argues. But wait, there is another mechanism! These green spaces provide a biodiverse microbial environment that our immune systems otherwise lack.[15]

He notes that exposure to natural environments seems to have more benefit for the poor, presumably because the rich are more likely to travel on holiday, keep a second home in the country, or maybe just play more golf and take their kids to the petting zoo on weekends.

All this is still partly speculative, but Rook hopes it will influence urban planners to work with ecologists and public health policy makers to rethink our cities. That would be a pleasant result for thinking about the complications of the immune system prompted by contemplating the equally daunting complexities of the microbiome.

Complex they may be, but the microbiome's influences on health, mediated by immunity, mainly involve just one of the body's key systems. But a look at conditions that could in theory be linked to the microbiota would not be complete without a glance at some other, still speculative but fascinating connections. They link our gut microbes with the most complicated organ of all, the brain.

microbiome influence more than our metabolism, our immune system and our hormones. The microbes affect our brains, too.

9 | Gut feelings

A mouse sits on a narrow, four-pronged platform, where it can explore the arms of a cross. Two ways are open, while two have enclosed sides, and the whole thing is a few feet off the ground. When it moves, it is fairly easy to time how often the mouse goes down one of the open tracks, how often it keeps to the enclosures.

This is one of the laboratory set-ups that is used to measure a mouse's mental state. Small rodents are rubbish at filling in questionnaires, but their behaviour can be measured. Shunning the open spaces is taken as an indicator of murine anxiety, while a bold mouse does not mind if the elevated track has walls or not.

Unexpectedly, you will find this rig in some labs where microbiome experiments are in progress. It turns out that the state of the microbial population in the mouse's guts can affect its behaviour. It is one of a collection of lines of evidence that our microbiota influence more than our metabolism, our immune system and our hormones. The microbes affect our brains, too.

Meet your second brain

There are already researchers speculating about using bacteria to modify our mood. Imagine your new breakfast choices in

2050. Yesterday didn't go so well, and you've had a restless night. You are feeling really pretty down this morning. So you ladle an extra dose of 'Inner Sunshine™' on your cereal to lift your mood. Thank heavens for modern probiotics. It must have been awful to have been around when all people had to get them going first thing was a double espresso.

Is this speculative vignette remotely plausible? It is easy to see that the mass of microbes in the colon can contribute to digestion. But how do their effects on other organs extend to the brain?

Brain and belly certainly communicate. Ask anyone who has felt their bowels loosen when faced with extreme danger. Those of us with quieter lives still acknowledge gut feelings. We know stress leads to digestive disquiet, that mood and appetite are linked. So perhaps the butterflies in my tummy can be set aflutter by the bacteria.

The possibility of bacterial populations in the gut influencing the brain is strengthened by the close connections between gut and nervous system. In fact, the gut has so many neurons of its own it has been dubbed our 'second brain'. It is, amazingly, an independent centre of neural activity, and this serves as a reminder of just how important the gut is. If you cut off other organs from the nervous system, they no longer function. The gut, as long as it is supplied with food, goes on doing its job, regulating digestion, for example, through the peristaltic contractions that move partially digested food along the intestine.

At the same time, the gut does also have a hefty direct connection to the real brain via a branch of the vagus nerve. This is one of the trunk roads of the nervous system, and connects the brainstem to the heart and lungs as well as the gut. It carries information in both directions, and its many functions have been studied exhaustively.

But while all this traffic between gut and brain seems to increase the possibility that gut bacteria and the brain can talk to each other, it does not make it any easier to pin down any specific effects.

A dip into medical history also increases my wariness about strong claims of bacterial gut–brain interaction. They seem to hark back to a short-lived fashion of the early 1900s for ascribing all kinds of disorders, physical and mental, to the then newly discovered bacterial contents of the colon. The idea took off from the germ theory. Bacteria were mostly bad. The ones in your colon were probably bad for you too, doctors argued, giving off nasties that produced adverse effects – a process given the scientific-sounding name autointoxication or intestinal toxaemia. Scottish surgeon Sir Arbuthnot Lane, a man whom science writer Mary Roach describes as 'a raging coprophobe', regarded the colon as the body's sewer, and helped boost what with hindsight seems a bizarre enthusiasm for removing big chunks of the colon surgically. This enthusiasm persisted even though the operations killed alarmingly high numbers of patients. Probably the worst outcomes recorded were the 75 deaths out of 250 patients whose colon was surgically assaulted in one US mental hospital in just three years after 1919.

The theory, if that is not too strong a word, of autointoxication also encouraged the still-surviving – and equally bizarre – application of colonic irrigation. Have you tried it? I have not. It augments the normal procedure for emptying the last part of your colon, going to the toilet, by squirting water up your backside to dislodge supposedly harmful deposits that cling on.

The notion of autointoxication also helped to sell

early probiotics, starting with yoghurt, promoted by Elie Metchnikoff who believed that *Lactobacilli* were the best kind of bacteria to have in your gut.[1] As he told readers of the popular American magazine *Cosmopolitan* (no relation to the current title of that name) in 1912, 'we fight microbe with microbe'. Before long, there were probiotic potions and tablets on the market promising a 'scientific' aid to good health. They were especially helpful, punters were assured, for avoiding neurasthenia, the catch-all diagnostic label then used for a clutch of nervous and depressive symptoms. The whole pitch strongly resembles the way alternative and complementary medicines and health foods still often get sold today. Make some not very specific claims about vaguely defined conditions whose supposed symptoms most people can identify with some of the time, then watch the cash roll in.

Serious tests invariably failed to detect the claimed benefits, and the use of probiotics as a kind of mental tonic faded out after the 1930s. However, one of the more unexpected outcomes of the new views of our gut microbiome is that a new collection of links with the brain and nervous system have begun to emerge. Perhaps, just perhaps, mood-altering probiotics could be a viable alternative to psychoactive drugs one day.

Well connected

Think of the four great communication systems the body relies on. It turns out that the gut, and the microbes within, can influence the brain through each of them. There is the important network of direct nervous system connections between the gut and the brain. In the endocrine system, passing messages via chemicals, bacterial signalling molecules also overlap with our own large catalogue of signalling molecules, including some hormones and neurotransmitters. Some of the hormones go on to affect gene expression, bringing in system number three. The general influence of the microbiota on the fourth network, the immune system, and its reactions and inflammation, can have secondary effects almost anywhere, including the brain.

One place looks particularly promising to those seeking out specific effects in this whole complex of interactions. We know that various strains of gut bacteria make a whole collection of signalling molecules that have roles in the body – including neurotransmitters like gamma-amino butyric acid (GABA), acetylcholine, and dopamine. Some have been found in higher quantities in mice with active microbiomes than in their germ-free counterparts, at least in the gut, suggesting that the bacteria are contributing to the level of these influential molecules.

Mark Lyte of Texas Tech University has argued for a couple of decades that this calls for creation of a new multidisciplinary subject he calls bacterial endocrinology. Given the lengthy history of microbiology it is surprising how recently this got anybody's attention. Lyte likes to tell the story of the first talk he gave on the subject, at the American Society of Microbiology's General Meeting in 1992. He began with an audience of two. One left soon after the start, leaving a sole listener – his lab technician.[2]

More recently, as the new findings about the microbiota have piled up and his ideas have gained an audience, he has emphasised that it implies two-way communication between host and microbes. As he puts it, 'a microbial organ within the gut exists in which bacteria communicate not only with the host, but also each other through the production and recognition of neuroendocrine hormones which have a long shared evolutionary history'.[3]

So far there are only hints, coming from animal studies, that this overlap in molecules and functions has discernible effects on brains. Results are extra hard to interpret because you have to believe inferences from things mice and rats do that are supposed to match states of mind in humans. Consider the moderately repellent 'forced swimming' test. Forced swimming because it involves dumping a rat or mouse into a water-filled cylinder it cannot get out of. If the beast gives up trying to stay afloat by frantically swimming, and becomes still, this is recorded as 'depressive like behaviour'.

Still, there are studies in which rats fed on soy milk enriched with GABA swam more determinedly than a control group fed ordinary milk. They matched the efforts of other animals given the Prozac-like antidepressant drug fluoxetine. So perhaps a big dose of bacteria that make GABA would have the same effect.

Another complication, beyond trying to interpret the behaviour of small rodents in human terms, is the way these ancient signalling molecules have evolved multiple roles. So, to stick with GABA, this little amino acid has been employed in the nervous system, where it usually acts as a neurotransmitter. But it also has effects in the immune system, tending to damp down inflammation. So we need to establish whether gut-generated GABA's effects on the brain, if any, are produced via the nervous system, the immune system, or both. That will involve painstaking studies of mice or rats with microbiomes of known composition, with and without species of bacteria that can make specific neuroactive molecules. If the results are to lead to any convincing conclusions about cause and effect, the creatures' guts will have to be sampled and analysed chemically as well as microbiologically at the time any behavioural measurement is taken. Then experimenters may have a smoking gun to convince a doubting jury that the bugs made the rodent behave the way it did.

'A salad for me, and a burger and milkshake for my little friends'

The new evidence that microbes could influence our behaviour raises some other disconcerting possibilities. Just because it turns out to be OK to have a teeming mass of bacteria in your gut, there is no reason to suppose they are looking after our interests. They reproduce independently. Therefore, evolutionary theory dictates, they will do things for us only as long as it is to their advantage.

Joe Alcock of the University of New Mexico and colleagues argue that we know bacteria in the gut are affected by diet, and that different species are competing with one another. If they can find a way of influencing us to eat what they like, they will do better than other cohabiting species in the same gut that fail to do so.

That could happen, they suggest, by inducing cravings for particular foods – for fat or sugar, for example. They point to evidence that people who crave chocolate have different microbial metabolites in their urine from those who are indifferent to it – even when they have identical diets. That definitely looks suspicious, although it strikes one that if the diets stay identical the 'craving' cannot be that strong.*

---

* If they did induce us to eat more chocolate, the microbiota might be doing us a favour – which they themselves make possible. Some gut microbes can metabolise polyphenols in cocoa powder to produce smaller anti-inflammatory molecules that have a benign influence on the immune system. Chocolate is not *needed* for this, sadly. Blackberries and tea also supply the bacterial substrate. Courage (2014).

But we know of other, more sinister, manipulations by the microbiota. The most alarming is that *Toxoplasma gondii*, a nasty protozoan parasite, can wipe away rats' fear of cats. Cats are the species' main host, and the only one in which it can reproduce sexually, so it is in the parasite's interest for the rat to be eaten. It is not just a matter of no longer being wary of felines, either. Infected rats may become sexually aroused by the smell of cat urine.[4]

You may or may not be sexually aroused by cat pee, but there are undoubtedly some things you do that are more in the interest of some of your gut microbes than others. So it does seem likely that bacteria get involved in the signal traffic here, and encourage us to eat things they like, or just to eat more. The possible routes are varied, and the reasoning inventive. The most indirect example proposed by Alcock and co. is that the persistent crying of infant colic often goes with altered gut microbes. Colicky infants also get fed more. Perhaps the bacteria are signalling to the parents that they need more food?

Bacteria could even nudge us by influencing pain in the gut, by releasing toxins when their nutrient supply runs short. That seems a crude signal, but maybe some people try to quell stomach discomfort by snacking on foods that supply the missing nutrients. Mouse experiments also show differences in taste receptors for fat and sugar in germ-free mice, so rewards as well as punishments may be on offer at the behest of our gut bacteria.

All those mechanisms operate indirectly, but the neuroendocrine channels offer plenty of possibilities for direct influence on appetite, both for big helpings and for particular foods. For now, all this remains pretty much untested in people, but the authors lay out a series of predictions that could inform a research programme to flesh out these speculations. For example, they suggest that food preferences may be contagious, via a loop in which microbes that favour the food in question reinforce the appetite, and people who live together share their microbiota as well as their dining tables. Meanwhile, those involved in the debate about the cause of obesity can chew over their conclusion that 'Exerting self-control over eating choices may be partly a matter of suppressing microbial signals that originate in the gut'.[5]

Beyond appetite

There are now suggestions that the state of the gut microbiota can influence, as well as be influenced by, pretty well every problem of brain, nervous system or behaviour – including mood disorders, schizophrenia, autism, and chronic fatigue syndrome. The strongest case at the moment relates to the most common mental illnesses, depression and anxiety.

One line of evidence concerns the effects of lipopolysaccharide (LPS). As mentioned before, this is the label for a class of compounds that stick out of the cell walls of some bacteria, anchored by a lipid tail. The lipid portion can trigger a cascade of effects, many of them leading to activation of immune cells and release of molecules that promote inflammation.

LPS helps hold together the outer membrane of the bacteria that make it, so they make quite a lot. There may be as much as a gram of it in your gut at any one time. Microgram quantities can cause a violent response, so it is risky stuff to have around. Healthy folks have vanishingly small quantities in their blood, but if just a little more gets across the gut barrier it can trigger the low-level general inflammation that is associated with a whole catalogue of adverse health effects.

One gut–brain connection here is top down. Some animal experiments suggest that personal history may weaken resistance to the effects of LPS and other nasties. Mice that have been stressed out by the crude but doubtless effective technique of electrocuting their tails release more inflammatory signalling chemicals – cytokines – when given a load of LPS than their unmolested cage mates. Of course, we know that in people the gut can both respond to stress and create it. The symptoms associated with inflammatory bowel disorders are themselves pretty depressing, and exacerbate the stress that is partly responsible for the condition, in a vicious cycle. But we are now looking for more direct effects on the brain.

There are other suggestive experiments with mice, but they do not all point in the same direction. Early results indicated that germ-free mice experience higher stress. This went along with lower levels of a key chemical influencing neural development – brain-derived neurotrophic factor or BDNF – in the hippocampus, the small brain region known to be a centre of learning and memory.

However, there are equally carefully designed studies that show contrary effects, featuring, for example, mouse strains known to be genetically disposed to be shy and retiring – for mice – that can be coaxed into behaving less timidly by a dose of antibiotic, which also boosts their BDNF.[6] There are also studies with germ-free mice that are *less* stressed than controls with a normal gut microbiome and again had stronger production of BDNF.

The differences in behaviour in the last study persist when adult mice are furnished with a microbiome, so there is an implied influence on brain development when the mice are growing up.[7] On the other hand, the mouse studies also extend to microbiome transfer experiments that appear to carry behaviour with them. The timid strain – labelled BALB/c mice – can be transformed to resemble one that is normally much more exploratory by adding the appropriate gut microbiome to germ-free individuals. And, to round out this rather startling result, the bold strain get anxious when colonised by the complementary gut microbiome. How does this work? We don't know.

If there are demonstrable influences from the gut, and its microbes, on the brain, it is even easier to show the brain influencing the microbiota. Again, mouse experiments indicate that stress is a key ingredient, tending to reduce gut microbial diversity and promote that other effect that always gets measured and usually turns up, inflammation. That suggests that feedback loops could emerge. Stress-affected microbial populations may disrupt controls on inflammatory cytokines that also have effects on the brain – including, it seems, increasing the risk of depression.

Validating any of these ideas in people is much harder than doing experiments in mice. As we do not know what causes the mental illnesses of most interest, much research is limited to recording changes that seem to be associated, some of the time, with one condition or other, and may also be affected by gut microbiota. The changes in question include detecting antibodies to host cells in brain tissue, alterations in the gut epithelium that make it 'leaky' for small molecules and perhaps even some bacteria, and in the so-called blood–brain barrier. Add all those together, and you can easily spin scenarios in which an influence passes from bacteria to gut, to brain, to behaviour or mental state. And such scenarios certainly exist, for conditions as varied as schizophrenia, mood disorders, obsessive-compulsive

disorder, autism, attention-deficit disorder, anorexia, narco-
lepsy, and chronic fatigue syndrome. In general, though, it is
very hard to cash in the one big new thing we know – that as
well as a gut–brain axis there is a microbiome–gut–brain axis
– to establish firm links in any of these cases. I will look at just
one more, which is typical of current debates.

Creating a disturbance?

On the face of it, autism is a good candidate for finding useful
new results from microbiome studies. Although this increasingly common condition's origins are unknown it is generally viewed as a neurodevelopmental disorder. And it is also
well-established that children with autistic symptoms may
also have gastric troubles – being more likely to suffer from
both diarrhoea and constipation than other children, and to go
on to develop inflammatory bowel disease (it never rains ...).
Pinning down how strong that association is, however, is hard.
Different studies indicate an incidence of gastrointestinal troubles in autism ranging between 9 and 90 per cent.

Microbiome samples have duly shown up differences
between the gut flora of autistic and non-autistic donors, with
the former tending to have lower microbial diversity as well as a
different balance of species. And complementary mouse studies
have both linked similarly altered microbiota with 'leaky gut',
and used B. fragilis or Bacteroides thetaiotaomicron to correct it
and thereby relieve some autism-like symptoms. The researchers who report this, once again from Sarkis Mazmanian's group,
suggest that it could lead to probiotic therapy for humans.[8]

The changes that led to the model autistic behaviour in the
mice in these studies, though, were initiated by exposing their
mothers to a virus when they were pregnant, mimicking the

maternal flu infections that increase the incidence of autism in humans. And the bacteria added later did not make all the mousy symptoms disappear. Their relative dislike of social interaction remained after the microbiome alteration. So a pessimist would say this indicates an approach that could one day lead to relief of some, but not all, autistic symptoms in some as yet undefined subset of autistic people who have microbiomes altered in a particular way.

Autism is probably a collection of conditions, and each of them may have multiple possible causes. In humans, we need more research to establish whether differences in gut microbes are regularly detectable. A systematic review of existing human studies published at the end of 2013 found many of them wanting, featuring small sample sizes, lack of standardisation, and 'generally poor methodological quality'. [9] Getting anywhere, said the authors, needs 'larger studies of high quality'. At the same time, there will certainly be trials of probiotics, either formally designed ones that get published in the journals or personal ones by hopeful parents.

They may be encouraged by one study that wheeled out neuroscientists' current favourite piece of kit – the MRI scanner – to see if we can make the effects of a probiotic visible. This might be a good workaround given that ethical constraints on experiments on people are a little tighter than when researchers work with mice. An MRI scan is non-invasive, and tells you that *something* is happening in the brain – specifically that blood flow, and presumably metabolic activity, is increased in some areas compared with others.

What it means is largely informed guesswork. But if I read that someone has combined probiotics and MRI scans I still want to look. What happened when Kirsten Tillisch of the University of California at Los Angeles fed twelve women yoghurt containing four kinds of live bacteria twice a day for twelve weeks was this. Put them in a scanner and show them faces with angry or fearful expressions and a large collection of small brain regions seemed to be working less hard than in either of two control groups – one of women who had a non-fermented milk feed that tasted like yoghurt, and one who had none at all. The researchers say their scatter of affected brain areas adds up to 'activity reductions in brain regions belonging to a sensory brain network'.[10]

So this small study did find brain changes, and they happened only in people who fed on probiotics. That does seem to lead to the conclusion that, as the report says, probiotics can modulate brain activity. The word modulate is carefully chosen, of course, saying just enough but not too much. It means some kind of signal is altered, in some way, in a process that is not really specified. And that is about as far as we can get at the moment in terms of actual facts. In terms of the grand speculation – of necking probiotics that affect my mood, or that could even help prevent or treat depression or other conditions – there is nothing in all these results to suggest that it can never happen. But, although some of the animal results look very suggestive, clear, usable information for people is some years down the line.

The hard problem

I could go on at this point to write about all the other conditions apart from autism that get listed as possibly subject to the influence of microbiome–gut–brain links. The conclusion would be much the same in each case. There is definitely a research area here that is rich in promise, but it is really only just getting under way. We know a little, but not much. We understand less.

That is not surprising given that we are introducing another fantastically complicated organ into the picture. Philosophers of mind call the origin of consciousness in the brain the 'hard problem', denoting that it is something we cannot yet explain satisfactorily. Understanding the microbiome–gut–brain axis is another hard problem. Following what is happening in the interactions between our microbiome and the rest of the superorganism already calls for an understanding of biological systems that integrates many disciplines. Bringing in neuroscience, which is full of wonderful things but also of things we do not understand, makes everything much, much more complicated.

There is also the splendid further complication that the mind can affect the body, and probably vice-versa, even when there is nothing apparently going on. The many manifestations of the medical importance of this, under the general heading of the placebo effect, include the following intriguing result. Give people with irritable bowel syndrome a placebo pill and *tell them it is a placebo*. This was a subtle kind of telling, along the lines of a double bluff. Instead of giving some patients a placebo and telling them it was a real medicine, the experimenters told them it was inert, but similar to pills that had been shown in earlier studies to improve bowel symptoms through 'mind-body self healing processes'. The placebo effect still worked. Patients told they were getting fake medicine reported twice as much improvement in symptoms as a control group given no treatment.[11] At the moment it is an isolated finding. But how intriguing that a condition like this, one increasingly the focus of attention from microbiome researchers hoping to provide relief for patients by understanding gut bugs' effects, can be improved by basically just telling people they are going to feel better. It is a useful reminder of how complex chains of cause and effect between gut, mind and brain are likely to be.

As we continue to piece them together, there is scope for them to become more complex still. For there is another dimension to the microbiome that remains relatively unexplored: the viruses within us.

10 | **Viruses are us, too**

The art of making things visible in a mass of DNA sequence fragments that no one could previously discern is developing rapidly. In July 2014 a new analysis of the mass of data from human microbiome samples made news worldwide by revealing a hitherto completely unknown bacterial virus in human faeces.

The startling thing about this virus was how much of it there was about. It is found in three-quarters of the samples examined – from Europe, Korea and Japan. And it accounted for around a quarter of the viral DNA in some people's microbial collection.

The report, from a group at San Diego State University, brought home how little we know about some aspects of our microbiome. We have a reasonably good idea which fungi live on people, and how many eukaryotes. Over the last decade, we have illuminated much of the hidden depths of our bacterial populations. But we are only just beginning to appreciate what may turn out to be an equally important part of the picture, the viruses.

Our own dark matter

Jonathan Swift, the great satirist, knew a thing or two about biology back in the 18th century. In *On Poetry: a Rhapsody*, in a passage so appealing it became a nursery rhyme, he wrote:

> … naturalists observe, a flea
> has smaller fleas that on him prey;
> And these have smaller still to bite 'em,
> And so proceed ad infinitum.

At the level of cells, the smaller fleas are the scraps of life we call viruses. Bacteria have them too. And along with the flood of new information about bacteria there is a smaller stream of results that are beginning to add another complex layer to the microbiota – hence to our superorganism – and our understanding of how it behaves and how it has evolved.

Viruses are more parasitic than symbiotic, consisting of small genomes in a simple wrapper that depend on living cells to reproduce. The viruses that use bacteria to make more of themselves are called bacteriophages, or just phages. The name comes from the Greek *phagein*, to devour, but if you watch them under a microscope, their action isn't one of eating by engulfing; rather, phages insert themselves into the bacterial cells, then make those cells burst once full of new viral particles.

Like cellular life, viruses are everywhere, and in much larger numbers than we suspected. Recent revelations of the number of viruses in the sea have boosted the estimates. Until the late 20th century, everyone believed that viruses in the sea came from rivers and sewage outflows. Now we know that every litre of the ocean contains as many as 100 billion viruses. That means the total of viruses in the biosphere is 10 million times larger than the number of stars in the visible universe.[1] That adds up to quite a biomass. If you collected all the viruses in the oceans they would weigh as much as 75 million blue whales.[2]

A good biological rule of thumb is that any given habitat has roughly ten times as many phages as bacterial cells. That seems to hold for the human gut. Given the number of bacteria we support, that is a lot of viruses. An early hint of how varied they are came in 2003, when a single human faecal sample yielded over 1,300 viral genotypes, most of them coming from bacteriophages no one had seen before.

Getting that information out wasn't easy. We think a gram of faeces typically contains between 10 and 100 billion bacteriophages. This is another of those new estimates that makes me think how it must have felt at earlier points when science insisted the world extended far farther than we knew – like the discovery of geological 'deep time' that left the Victorians peeping over a chronological abyss into a past that all of a sudden extended for hundreds of millions of years. Now, we can peer over the lip of the toilet bowl and contemplate the likely presence of thousands of millions of bacteriophages that, until a few muscular contractions ago, were waiting for the chance to reproduce inside one of the bacteria dwelling in the colon.

Viruses take over the protein-manufacturing apparatus of host cells, so need no ribosomes. Hence there is no 16S viral RNA, and no quick way of identifying what types of virus or phage are present in a sample. Instead, researchers go on one of those more complicated trawls, sequencing millions of DNA fragments and trying to figure out which ones look like they came from viruses. As most of the viruses now showing up haven't been seen before, this is pretty hit and miss. Some studies clean up samples by removing bacterial and human cells and what appear to be non-viral DNA sequences before sequencing what is left. This probably loses quite a lot of viral DNA. An alternative, as usual, is to get those machines running and sequence *all* the DNA, which avoids loss but maximises the amount of sequence recorded with no known matches.

Our knowledge of the myriad viruses that share all the sites in the body where microbes live remains very patchy, though it is expanding fast. Simple sampling and sequencing studies produce lots of results that are hard to interpret when we don't have a good overview of the population. That overview will need to be dynamic, because viral populations, as well as the genes of individual viral species, can change fast. In these early days the indications are that our individual viromes differ even more than the bacterial populations we carry. Samples from different people can yield viral profiles with hardly any overlap, and it could turn out that there is nothing like a core virome – or a set of viruses or viral genes that is always likely to be there somewhere, whatever optional extras have muscled in. The extent of uncharted viral genetic depths has led some to talk about 'biological dark matter', like the dark matter cosmologists invoke to explain the gravitational pull of galaxies that appear to have too few stars for the force they are producing.

However, this dark matter is not invisible or undetectable, just unusually hard to get into view. Another metaphor was suggested by a group from the University of Brighton in 2013: there are abundant 'subliminal' viral sequences.[3] They are usually too faint a presence to notice, but they can still be analysed with the right tools. And there are a few general things about these viruses that researchers are pretty confident about from earlier studies. Some come from the extremely detailed dissection of individual bacteriophages, especially those that infect *E. coli*, which were long a mainstay of molecular biology. Small genomes and rapid replication make phage experiments fast and, sometimes, simple to do. The very first complete genome sequence, published in 1977, was the small, closed DNA loop of the phage phiX174. It has 5,300 bases that code for just eleven proteins. The same tiny genome has since been made artificially, by Craig Venter's lab in 2003, so its place in history is assured.

But we need to know how things work in an endless ocean of viruses, and that takes us beyond knowing how to assemble one phage. One vital aspect of this is what recent molecular genetics has revealed about how viruses, including the ones that live in us all the time, interact with other organisms.

Viruses make genes mobile

Bacteria and their many viruses have been co-evolving for billions of years. Each phage recognises a specific target on a bacterial surface, and usually targets only a single kind of bacterium. A bacterial species and its viruses continually make small alterations that affect this interaction – the phage rather faster than the bacterium as it reproduces faster and often has a higher mutation rate. Indeed, viruses seem to be intimately involved in evolution – not only of bacteria but of other organisms, too.

Phages have two main ways of using bacteria. They can enter a cell, switch the machinery already there to making millions of copies of themselves, then escape when the cell bursts and find another target. Or they can operate in 'stealth mode'. In this case, the viral gene sequence is copied into the host's DNA, and replicates along with the bacterial chromosome each time the microbe divides. It only reverts to plan A if the bacterium falls on hard times, threatening its reproduction. Then massive viral replication and cell death ensues, and the phage looks for a new home. Until then, the bacterium and phage co-exist. Our bacterial symbionts, in other words, have symbionts of their own.

This slightly spooky ability of viruses to stitch their DNA into host cells has left traces in the genomes of every organism we've looked at, not just bacteria but also eukaryotes. Retroviruses, an important class of viruses that include HIV, are 'retro' because they preserve their genetic information in DNA's molecular cousin, RNA, which can be copied into DNA by an enzyme called reverse transcriptase. So-called endogenous retroviruses are stretches of chromosomal DNA copied from viruses that normally carry their genes on RNA. They are found scattered throughout the human genome, and account for around 8 per cent of our total DNA. Some may not be of viral origin, however, as there is another crucial evolutionary mechanism that gives rise to a bit of genome that looks like this.

Lots of genes, it turns out, carry sequences that allow all or part of them to be copied and then spliced into a different location. These jumping genes, or transposons, do naturally what genetic engineers do in the lab, moving DNA from one place to another. In some cases, this works via an RNA intermediate, which is then copied back into DNA by reverse transcriptase – famous in the history of molecular biology for demolishing the then-established 'central dogma' that information only ever moves in one direction, from DNA to RNA to protein. We cannot replay the history of pieces of DNA evolving, but it seems very likely that some retroviruses originated like this: a piece of DNA was copied into RNA, then cut loose, ending up as a free-floating replicator in the biosphere's genetic soup.

All these gene movements, and the relics they leave behind, have transformed our understanding of species and evolution.

The simple conclusion is that genes move around between organisms much more easily, and more often than we thought.

These gene swaps began in and between bacteria and their many viruses. They are still going on all the time. The simple option of some metagenomic studies that treat the entire contents of the colon, for example, as a kind of genetic soup suddenly looks smart. When genetic exchange is so promiscuous it matters less which individuals, whether bacteria or viruses, are present.

Seek, and ye shall find

We are now finding viruses, free-floating or inside other cells' collection of genes, everywhere we look. There will be much more to come from the virome, but we can already see that these simplest players in the ecosystem have important interactions with all the rest. That is not to say that there are not some viruses that are ingested and simply pass through the human system without doing much of anything. One early study, in 2006, reported finding a billion RNA viruses in every gram of human faeces.* Most of them, though, were not long-term residents of the gut, but plant viruses passing through. The most common in this sample, known as the pepper mild mottle virus, was still infectious. You can eat a salad, get your digestive system to work on the vegetation, and pass on an intact virus that can still cause disease – in plants in this case. Innocuous? Not necessarily. A further study suggested that the virus was linked to cases of fever and abdominal pain, so

---

* By the way, not *all* viruses are small. A new range of 'megaviruses' has been discovered since the millennium. They are DNA viruses that live inside eukaryotic cells. These, too, have been found on occasion in human samples (Colson 2013).

what passes for a reaction to a spicy meal could be a reaction to a plant virus.[4]

If that particular plant virus does cause discomfort, it is probably a simple immune response. For longer-term viral residents in our microbiota, the kind of more complex trade-offs we find at other levels are constantly in play.

Viruses and bacteria are each trying to look after their own interests – most narrowly by changing in ways that maximise their chances of surviving and reproducing. Sometimes this leads them to compete, sometimes to co-operate. Either *may* affect our interests, for good or ill.

On the minus side, viruses often move genes from one bacterium to another. There is strong evidence that they transfer genes for antibiotic resistance between strains.

David Pride of Stanford University, who made early studies of samples taken from the mouth – where there are often 100 million viral particles in a millilitre of saliva – showed that the phages there had an unusually high incidence of genes for antibiotic resistance.[5] They act as a kind of reservoir for these genes, he suggests, as the half-dozen human subjects who gave him their saliva had not been given any antibiotics recently.

The phage–antibiotic resistance connection is reinforced by experiments with mice. Collect mouse droppings before and after they are treated with antibiotics, then sequence all the phage DNA you can find. The result shows that the drugs lead to phage genomes that are richer in antibiotic resistance genes. The ability of small organisms put under strong selection pressure to swap genes looks uncannily like intelligence. The post-treatment phage genomes are enriched in genes that help bacteria resist the drugs that have been administered. More surprisingly, they also have more of the genes that promote resistance to antibiotics that the mice, and their gut microbiome, have not actually encountered.

The sense of opening a new view on the complexities of response to our own blunt antibacterial instruments is reinforced by an additional finding. Drug treatment strengthened networks of gene exchange between bacteria and phages that involved a range of other genes affecting bacterial growth and colonisation. The bacteria, it seems, can raid phage genomes for defence aids when they come under attack, and use the phages as a kind of genetic repository that can be drawn on to make their feeding and metabolism more efficient.[6]

On the other hand, the right bacteriophages may themselves be more refined instruments for attacking bacteria. There have been sporadic efforts to use preparations of phages as antibacterial agents ever since they were discovered. They faded away after the advent of antibiotics, but are getting increased attention again now that resistance to the drugs is so widespread.

Bacteria, meanwhile, have been susceptible to phages all along. And our own internal accommodations with bacteria make use of that. The most interesting finding so far about phages in our gut is where they tend to concentrate. There are lots more in the mucus that coats the gut epithelium than elsewhere. Close study of viral genomes shows why. They can bind to components of mucus.

The phages that dwell there are happy to co-exist with their host bacterial strain, reproducing along with the bacterium by integrating into its genome. But they attack bacteria that compete with their host. From our point of view, this probably inhibits bacterial colonisation of the mucus-coated surface.[7] Simple experiments with a single strain of phage support this idea. The effect could be adaptable to changing bacterial populations. The phage proteins have parts that constantly vary as their genes mutate, and mucus turns over rapidly. Here a product of the competition between phages and bacterial strains has led to a co-evolved niche that helps strengthen the human host's ability to keep bacteria in check.

There is also accumulating evidence of phages being brought into contests between bacteria. *Enterococcus faecalis*, for example, is a common gut bacterium that occurs in a wide variety of strains. Like others, it can incorporate phages into its genome. And strains that harbour phages can release them to reduce the population of their competitors when they are colonising germ-free mice. Working in concert with the phage gives them an edge in monopolising the nutrient-rich environment they suddenly find themselves in.[8]

These kind of interactions between populations, and sub-populations, of bacteria and phages must be going on all the time in the gut, especially in the densely inhabited colon. Most have remained inscrutable until now. As the discovery of a completely new but ubiquitous phage underlines, we still do not know all the phages that can be found in the microbiome, let alone what effects they have. The phage in question may itself have indirect effects on human hosts, but that speculation remains to be tested.

The novel phage that caught the world's attention in 2014 was identified by combing through DNA sequences to look for viruses using a new computer technique. The cunning feature of the program the researchers used was that it did three things at once. It highlighted DNA fragments that had a high probability of coming from a virus, then joined them together using overlapping bits of sequence until a complete, unknown phage genome appeared. But a phage without a bacterium is just a homeless scrap of DNA hiding in a coat of protein. The analysis also showed which bacterial DNA types the phage fragments tended to show up with.

That allowed the researchers to infer that their new phage, dubbed crAssphage, for 'cross-assembly' – the computer technique that revealed its genome – infects a *Bacteroides* species. That is one of the bacterial types that is usually abundant in the gut, and that plays a role in many studies of obesity. A phage that affects the number of *Bacteroides*, or even allows them to acquire new genes, could thus prove to be involved in all manner of far-reaching metabolic or immunological events.

Evolving now, in your very own intestine

Bringing viruses and phages into proper focus in the big picture of the human microbiome is a task for future researchers – the preliminary studies show how large a job there is to do.

However, if you grant viruses the status of living things, then virome studies have already established one more startling thing. There are new species evolving in your gut. The pioneering studies led by a post-doc student, Samuel Minot, in Frederic Bushman's lab at the University of Pennsylvania included a small effort to assess how stable one person's virome might be by following one volunteer stool donor for two and a half years.

The results of regular DNA analysis of viral genes showed that most of the viral types that lived in their subject's gut at the outset were still there 30 months later. And, while viruses reproduce rapidly and mutate fast, most of them stayed more or less the same. Most, but not all. One family of viruses, the *Microviridae*, which have a short, single-stranded length of DNA to pass on their genes, mutated often, and the DNA sequence changed by as much as 4 per cent over the two and a half years. Some already classified *Microviridae* species differ by only just over 3 per cent, so it seems fair to regard this as equivalent to speciation. That is something worth pondering for anyone interested in biology. The origin of species, the great phenomenon that Darwin spent so many years pondering, is going on inside you. New species, albeit on the smallest possible scale, are appearing in your gut, and presumably other microbiome sites, all the time.[9]

Speedy evolution at this level has been going on as long as humans have been around, indeed far longer. Our own pace of evolution has been much slower, to the point where there is sometimes speculation that human evolution has come to a halt. But we, uniquely, have added cultural evolution as a force for change in the natural as well as social worlds. As we've seen, that has brought important, unsought changes in our microbiome. Suppose, though, we begin to alter it deliberately, perhaps for the better? Time now to consider where cultural evolution could take our superorganism in future.

11 | Civilising the microbiome

The human microbiome is poised for an epochal shift in how it is established and maintained. It evolved with us. Like other organisms that accommodate microbes, we have developed ways of guiding the right microbes to become colonists and discouraging all the rest. Since we began eating fermented foods, the human microbiome has also been influenced by culture, (in both senses!). As we have seen, in the last 100 years the advent of home hygiene, modern diets, caesarean births and mass use of antibiotics all mean that our cultural impact on our microbiome, largely unwitting, has become much greater.

Now we are poised to adopt a more informed approach, in place of the general onslaught we unleashed for what seemed like good medical reasons, and the inadvertent impact of our eating habits. There is a long way to go. We are like agriculturalists who have just discovered slash-and-burn clearance contemplating the path to fine horticulture. But it seems logical that knowing more about our microbiome should lead to ways of improving it. This time, what counts as improvement will be decided by us, not the microbes.

Better to give than receive: faecal transplants

The crudest approach, in several senses, is to give someone the microbiome they may need by transplanting it wholesale. Some bowel conditions are now being treated this way. This at least gives us a kind of baseline – it is the procedure you adopt when you do not really know what you are doing. As it involves using a donor microbiome in the form of faeces, it evokes mixed feelings. We are simultaneously fascinated and repelled by shit.*

It *is* a desperate measure. But this crude procedure does have something to offer. It may be a fit treatment for damage caused by the almost-but-not-quite equally crude approach of waging chemical warfare on microbes in our bodies.

The people who definitely benefit from refreshing their colon with someone else's faeces have a persistent overgrowth of the common bacterium *Clostridium difficile* in their guts. This can arise as a result of antibiotic treatment for some other condition, which kills off the mixed population that previously competed with *C. difficile* and kept it at tolerable levels. In this sense, the problem is what doctors call an opportunistic infection.

Chronic *C. difficile* infection in the colon is truly dreadful. The patient is afflicted with painful cramps, nausea and dehydration, while suffering foul-smelling diarrhoea. Rapid weight loss follows and in serious cases, in those unfortunate enough to have the worst strains of the bacterium, the bowel inflammation can even be fatal.

Normal-looking faeces are conspicuous by their absence,

---

* While we are being fascinated, repelled, or even amused by excrement, we might also remember that separating it from our lives is a recent achievement, not yet complete. An estimated 1.8 billion people still rely on drinking water contaminated by human faeces. Bain (2014)

and treating *C. difficile* infection with the contents of a well-functioning bowel is not a new idea. Various historical precedents have been cited, going back to Chinese accounts of faecal tea 700 years ago, but the first modern medical publication dates from 1958[1] (it didn't catch on). A similar approach has also been used in veterinary medicine to treat horses with digestive problems and cows with mastitis. Proponents also note that many species, from elephants to chimps, eat one another's faeces – laboratory mice share their gut microbiomes this way unless researchers take special care to combat their habitual coprophagia. It seems that declining to snack on our conspecifics' excreta is another trait that marks out humans from most other mammalian species. The recent revelations about the complexity of our gut microbiome, along with continuing frustration in treating *C. difficile* patients, have encouraged a new era of experimentation with stool.

The procedure is simple and, as doctors who do it attest, pretty unpleasant to administer. You can think of it as transplanting a whole complex ecosystem but that is not the first thing that comes to mind when you take healthy donor shit, liquidise it in a blender, and pass it into the colon via a nasogastric tube, an enema or a colonoscope, depending on who does it.* There are alternatives, such as capturing the transplant material in gelatinous capsules that can be swallowed, but you have to make sure these pass through the stomach – which kills many microbes – so the new inoculation reaches its intended destination in the large intestine. The quantities are not small. A typical procedure might involve the contents of half a dozen

---

* A 2012 patient survey found that receiving faeces by nasogastric tube was the 'most unappealing' method on offer. Zipursky (2012)

50ml syringes. One preliminary report does suggest that frozen samples, in capsules that can be swallowed, yield promising results.[2]

Most data so far arise from use of other, cruder methods and, for all the unpleasantness, the procedure does appear to work. Accurate figures are hard to come by, as the combination of publicity, simplicity, and the unbeatable motivation of an epically uncomfortable bowel condition have led to quite a few people trying it at home – as a visit to YouTube will confirm. Home equipment includes latex gloves, a kitchen strainer and blender, saline mix and enema bags, and, we assume, lots of disinfectant. Everyone agrees that a screened and standardised donation would be better, avoiding risks like hepatitis or even HIV from unscreened donors, but at the moment demand outstrips supply. That may change, as commercial providers get going, but faecal transplants pose tricky issues for a regulatory system that is geared to deal with drugs or foodstuffs, not preparations intended for medical use but which can only be classified as 'neither of the above'.[3]

Meanwhile, the trials that have been reported in medical journals do indicate some very good results in patients with severe *C. difficile* infections that were untreatable any other way. A large majority, as many as 90 per cent, of transplant recipients treated for *C. difficile* improve quickly and the effects are often long lasting. One trial in the Netherlands in 2013 was stopped because the faecal treatment group fared so much better than *C. difficile* patients given the antibiotic vancomycin, the researchers thought it unethical to carry on with a control group.[4] More detailed follow-up shows that the gut microbiome community shifts to resemble that of the donor, and the alteration is usually stable.[5]

There is generally less or no evidence that transplants are any good for other conditions that have been suggested as possible targets, though some trials are in progress. The list of ailments for which there are either trials or case series published runs to irritable bowel syndrome, chronic constipation, ulcerative colitis, Crohn's disease, metabolic syndrome, chronic fatigue syndrome, multiple sclerosis and autism. Inflammatory bowel disorders do look like possible candidates for the treatment, but for the others it remains a long shot.[6]

Patient demand will ensure the procedure gets used more widely. Transplant providers are appearing who screen and store donor samples that are shipped to doctors on request. In the US they include the non-profit OpenBiome, founded by a group at MIT who were impressed by the results of a friend's faecal transplant.[7]

Viewed askance, though, it is a dumb treatment, or at least one born of ignorance, like trying to improve a disease-ravaged, depleted shrubbery by transplanting an entire rain-forest all in one go. The interesting developments will come when we learn how to get smarter about cultivating our microbial gardens. Plenty of people are convinced that is already possible, by the less direct route of eating the right microbes – probiotics. But do we know enough yet for that to do much good?

From faeces to food

The point of probiotics is to fine-tune the microbiome by starting at the point of ingestion rather than excretion – obviously more appealing, then. Probiotic products, usually involving just one or a few microbes, are many and varied, and are already the basis of a multi-billion-dollar global industry.[8] The idea has history, but has had a big boost from contemporary studies of the microbiome. Supermarkets and health food stores offer many probiotics. People who take health maintenance seriously often believe that good bacteria can help. But what exactly can probiotics do, and are they ready for more carefully tailored applications?

There is one sense in which we know probiotics work and are, if not essential, then definitely beneficial. The realisation that breast milk furnishes a feeding infant with microbes makes it the first probiotic we can encounter. In fact, as it also contains food for bacteria that are destined to colonise the infant gut (as discussed in Chapter 5) it is a combined probiotic and prebiotic, the term for foodstuffs that encourage bacteria to grow rather than supplying the actual microorganisms.

We can add that to the plus points for breastfeeding, already recognised as preferable to formula milk. But assessing the case for probiotics beyond this stage of life is not straightforward. There are strong advocates for dosing yourself with the right bacteria, either for general health or to help prevent or relieve some specific condition. Often, though not always, they have something to sell. The big companies marketing probiotics, such as Danone, are also strong supporters of microbiome research, especially when it involves probiotics. This does not necessarily skew the researchers' results but the companies obviously hope that some of them can be glossed in ways that help shift more product.

Advocacy evokes scepticism, and there are also people keen to shoot down claims for the benefits of probiotics. The history here is long and the arguments tangled, but it is worth bearing in mind that probiotic enthusiasts often set great store by the fact that the weight of tradition appears to be on their side. Eating fermented foods is certainly very widespread – it has been historically and it is in many parts of the world today. That is taken to mean that they offer an essential contribution to our diet, and that it is microbial. A decline in consumption is then regarded as part of the modern onslaught on our microbiota, and it is suggested that eating more of them will help 'heal' your microbiome, or possibly your gut, by restoring 'balance'.*

There actually is a modern onslaught on our microbiome, but failing to eat enough pickled cabbage or yoghurt may not be part of it. The few surviving hunter-gatherers and, presumably, their numerous forebears, do not eat fermented foods.[9] Anyone who wants to cultivate an ancestral microbiome would have to go the whole way and embrace a Palaeolithic diet.** Once humans shifted to agriculture, farming societies needed to find ways of preventing spoilage if they wanted to store food – after harvest, say – in an environment where microbes are everywhere. The best way to do this, often the only way, is to ensure the food is colonised by microbes that change it but

---

* These are the characteristically vague terms used repeatedly by, er, pro-probiotic writers, such as Raphael Kellman, MD, author of *The Microbiome Diet* (2014). We certainly don't know enough yet to prescribe any such thing, a major reason why – I must confess – I have not made time to read it. However, one of his articles proposes a diet that includes, among many other items, hydrochloric acid as one of the things we may need 'to replace stomach acid and digestive enzymes'. I have absolutely no clue what this might have to do with the microbiome but it is a seriously bad idea. See http://www.everydayhealth.com/columns/health-answers/balanced-microbiome-key-health-weight-loss/
** This may even resemble the 'Palaeo diet' now promoted in some quarters, but the anthropological evidence is that the real thing would involve a lot less meat and a lot more chewing leaves than most moderns would tolerate.

leave it edible. As well as being delicious, whether that judgment is innate or learnt, foods from cheese to kimchi to dosa to bean curd keep longer than their unfermented equivalents. The modern, industrial diet may simply contain fewer fermented foods because we have now invented canning and refrigeration. Whether particular probiotics do any other good is a question best decided by research that takes off from what we now know about the gut microbiome, not untested generalisations based on the fact that lots of people have eaten them in the past.

In most cases, there is not enough research yet to be sure either way, so we are all free just to go on eating the ones we like. There are some striking results from experiments with probiotics, but they do not yet translate into a case for regular use for microbiome maintenance, or tell us what kinds might be best for that, one day.

Bacteria for the 'glow of health'?

Here's the kind of finding that could send you down to the health food store: 'It was recently shown that feeding of probiotic bacteria to aged mice rapidly induced youthful vitality characterised by thick lustrous skin and hair, and enhanced reproductive fitness.' Sounds good, where can I get some please?

This is no advertisement. The words come from the abstract of a scientific paper.[10] The lead author, Susan Erdman of MIT, is the closest I have come across to an out-and-out scientific advocate of probiotics, although her enthusiasm for their possibilities is built on a lengthy research career investigating inflammatory responses and cancer. Out of that work came an interest in bacterial influences on immune cells, and a series of experiments in which mice dined on yoghurt laced with a well-known probiotic bacterial strain, *Lactobacillus reuteri*.

The results were striking. Compared with mice on a regular laboratory diet, the probiotic eaters had a 'glow of health', with thicker, shinier fur. When the experimenters shaved the backs of some mice, a few days later the probiotic group had regrown their fur while the control group still sported a patch of naked skin. These visible effects went along with other changes including increase in skin and mucosal thickness and adjustments in control of inflammation. The effect was subtle – the probiotic's influence did not show much in the way of detectable effects on microbial ecology. However, in Erdman's summary, the probiotic 'is reprogramming the host's entire body'. When she described these experiments at a conference in Cambridge in the UK in a presentation full of pictures of disgustingly healthy-looking mice, the question everyone plainly wanted answered was not a properly scientific query, so there was a small queue of people waiting to ask her over coffee afterwards: does she now take *L. reuteri* herself? Yes, yes she does.

If I had done experiments like these, I suspect I would, too. They became more persuasive when the work was extended by injecting a key class of T lymphocytes taken from probiotic-dosed mice into normally fed mice with no lymphocytes. They responded as if they had benefitted from the probiotic, with the same plumped-up skin and glossy fur.

What to make of this? The mice do seem to benefit from this well-defined probiotic. On the other hand, the work was done in a lab very well-versed in the molecular and cell biology assays needed to get a clearer picture of what the effects depended on. The ancillary experiments point to a complex set of interrelated controls on the immune system and on hormonal regulation (involving oxytocin) as the keys to the underlying mechanism. Is it likely that this bacterium is the only way this system achieves the state that makes the mice healthier in this handily visible way? I imagine not. Would it work in people? We do not know yet. If you are already pretty healthy, would it make any difference? Again, we do not know.

And the big question. Is it a good idea to experiment with *L. reuteri*, which is widely available in health food shops? (Among other things, it is marketed as Cardioviva, supposed to act as a cholesterol-lowering aid to heart health, so it seems a pretty versatile strain.) I am not sure. Many people, perhaps most, already have *L reuteri* in their guts anyway. It is a normal inhabitant of the intestine. It is even found in breast milk so babies often get an early dose naturally. Probiotic capsule users do post enthusiastic reviews, but generally just say that it helps digestion and keeps them 'regular'. No mention of unbeatably attractive skin and hair. I am regular already, thanks, so I think I'll leave it for now.

The link with such spectacular results, albeit in mice, does make *L. reuteri* stand out from other probiotics, though. So does the fact that you take it in a capsule, which protects it on the way through the stomach, and that it probably contains the strain indicated on the label. Not many probiotics are like that. The majority of the products out there are just trading off a general impression that probiotics, or simply fermented foodstuffs, are good for you. Go to the website for a maker of big-selling products like Danone, for example, and there are lots of references to good digestion, microbial 'balance' and 'immune health' that can be achieved by regular consumption of Actimel yoghurt (branded DanActive in North America) – but little in the way of actual science. This may be related to legal action pursued against the company soon after the product was launched, when it made more specific claims that were widely criticised. Now it mainly wants you to know that the stuff contains a patented live culture of *Lactobacillus casei*, in addition to the regular yoghurt-making bacteria *Lactobacillus delbrueckii* and *Streptococcus thermophilus*. You can get the last two from pretty much any live yoghurt you eat. Do the additional benefits of the probiotic strain they highlight justify paying a bit more? I am not really persuaded.

That makes it fairly typical of currently available probiotics. Most commercially available ones still use a small number of strains that predate our detailed knowledge of the microbiome, their selection going back to the 1930s or even to the theories of Elie Metchnikoff in the early 1900s. They are almost all *Lactobacilli* or *Bifidobacteria*. These bacteria do not normally live in large amounts in the gut. They do tend to live on the intestinal mucosa, though, so it is possible they may have more influence on immune regulation than their overall prevalence would suggest.

Their long history of use assures that they do no harm. Sadly there is little evidence they do good. Industry likes them because they are easy to culture, survive well in acidic products like yoghurt, and are not easily damaged during food processing. Whether our digestive system benefits from them is another matter. You can probably do yourself as much good by choosing a microbiome-friendly diet. That is pretty easy, in one sense, as it is much the same as the recommendations for a healthy diet we all know already, even if adhering to them is hard: avoid processed food, do not eat too much meat, and have lots of fresh fruit and vegetables. A few, including greens, beans and artichokes, may be particularly effective at helping maintain a microbiome that does its job well. And there may be a case for making sure you eat some fermented foods, whether that means yoghurt, cheese, kimchi or sauerkraut.

Meanwhile, wait for more research on different potential probiotic species. We now know how many other species live in our intestinal tract, and there must be as many more possibilities for probiotic use of different bacteria, provided they can be grown. They need careful testing, both in the obvious sense that trials need to be large enough to detect any beneficial effects and have proper control groups, but also with care in specifying exactly what is being tested.

A good example is a probiotic that does have some well-attested benefits. In *Good Germs, Bad Germs*, Jessica Snyder Sachs rightly singles out a commercially successful strain of *Lactobacillus rhamnosus*, first isolated from human gut samples in 1985. It is known as *L. rhamnosus GG* after its two discoverers, Sherwood Gorbach and Barry Goldin of Tufts University, who hoped that gut isolates would be a better source of probiotics than strains found in dairy products. And so it proved, with this new isolate turning out to be an effective coloniser and a useful aid to prevention of diarrhoea and inflammation caused by gastroenteritis. Sachs, writing in 2007, reckoned it was 'the most thoroughly studied of modern-day probiotics', a fact the vendors of the commercialised preparation, marketed as Culturelle, understandably feature in their advertising.

Bacteria are full of surprises. In 2009 Matti Kankainen and colleagues at the University of Helsinki compared the genome of this strain with similar bacteria. They found that *Lactobacillus rhamnosus GG* carries a previously unsuspected cluster of genes that help it make long, thin projections poking out from its cell wall.[11] These 'pili', made mainly of protein and with adhesion molecules on the end, had previously been seen only in pathogenic bacterial strains, which use them as an aid to infection. In *L. rhamnosus GG*, experiments suggest they are important for helping the microbes bind to intestinal mucus, and in stimulating the host immune system. We also know now that mutants which do not make the pili exist, and that the structures are relatively fragile. It is rather easy to knock them off. So providers of commercial products should be testing them to see if these vital appendages are present on the bacteria they are growing, or whether they survive processing. Their structure and properties are still being worked out. Perhaps one day a product that provides purified components of the pili will be on offer instead.

The optimal probiotic will also be harder to specify because potential benefits will probably come from combinations of bacteria. The precedent for likely future products comes from livestock, specifically chickens. Research on preventing *Salmonella* colonisation in chickens – which makes it easier to keep it out of eggs – suggested that encouraging other bacteria normally found in chickens to occupy their guts when young helped stop the one that causes food poisoning getting going. One result: PREEMPT™, a preparation of 29 bacterial strains licensed by the US Food and Drug Administration back in 1998 as an early treat for baby chicks. Feed them pellets or even spray them with the bacterial soup and their intestines end up occupied by healthy bacteria, not nasty ones.

There are companies working on similar formulations for other farm animals, and they see no reason why the principle should not apply to people. Andrew Serazin of US biotech company Matatu suggests thinking of the microbiome as 'a kind of software' that can be upgraded or patched. The early bacterial colonists that make their home in the infant gut are 'a developmental resource', he says. The big question then is, 'Can early childhood inoculation with a defined consortium of microbes ensure appropriate development of this system?'

That might point towards a future in which parents feed their infant a cocktail of designer probiotics. Commercialising a product like that faces a few obstacles. If the inoculation is made up of bacterial strains isolated from people – presumably healthy ones – the patents that biotech entrepreneurs like to help them move into profit will be difficult, perhaps impossible to write. Even if one were granted, there are so many similar bacterial strains up for grabs it would not take much microbiological nous to work round it. Controlling the production of a preparation with perhaps dozens of bacterial species would be a nightmare, but vital for regulatory approval. And public acceptance of simple probiotics, mainly in food, is no guarantee that people will take to the idea of stuffing baby full of bacteria to help him or her grow up strong and healthy. Expect to see Serazin's company, and others, concentrate on agricultural applications for now.

Back to the ecosystem

The deeper you inquire into probiotics and the prospects for inoculation with useful bacteria, the more it feels like most of the work of teasing out what might deliver reliable benefit remains to be done. This comes across when looking at use of probiotics in patients where the benefit ought to be obvious. Examples where real benefits have been claimed after large-scale studies remain controversial. Whenever there are multiple trials, there are mixed results, and the overall conclusion of surveying lots of studies depends more than is reassuring on the ones chosen for review.[12]

The best-known example is the use of probiotics to help prevent *C. difficile*-associated diarrhoea in patients given antibiotics. Systematic reviews, such as the one published by the UK's authoritative Cochrane Collaboration in 2013, generally conclude that probiotics can do some good. That review covered 23 randomised controlled trials and found that among adults and children taking antibiotics for *C. difficile* infection 5.5 per cent overall still got *C. difficile*-related diarrhoea. With probiotics, that figure came down to 2 per cent on average.

That still does not convince everyone, however. The studies on offer may not be sensibly bundled together. They involve different probiotic strains of bacteria, and different treatment regimes. And a subsequent, large-scale trial in South Wales and North-East England involving 3,000 elderly patients found little or no difference in the incidence of *C. difficile*-induced diarrhoea in those given a mixed-strain probiotic.[13] The current best prescription? Once again it is to wait for results of more, well-designed trials.

Which brings us back to the smaller group of patients suffering the very worst effects of *C. difficile*. The advent of faecal transplant for cases that do not improve with other treatments marks the opposite extreme from simple probiotic use – the whole damn ecosystem. And it is investigating the gut microbiome as an ecosystem that will bring some simple improvements.

That is partly because a faecal transplant is such a crude, if effective, approach that from some points of view almost anything would count as an improvement. But it is also because the gut microbiome is an ecosystem that can be maintained in something like its normal state in the lab, if you know what you are doing.

Someone who does know is Emma Allen-Varcoe of the University of Guelph in Ontario. She begins from the position that quite a few conditions are linked with a disturbance of the gut microbial population. The medical and scientific label, as I mentioned in Chapter 8, is 'dysbiosis'. Not particularly helpful, says Allen-Varcoe, because 'nobody really knows what it is. We need to look inside the black box.'

The way to do that is to build your own box, or one quite like it. Her lab runs a model of the lower parts of the gut ecosystem in a series of linked chemostats – christened Robogut. The set-up is a line of carefully monitored flasks, each feeding into the next and supplied with suitable nutrients. The whole rig covers six feet or so of lab bench and, she says, 'can support the whole gut microbial ecosystem for several weeks at a time'.

The final output, as you might guess, is an unattractive brown fluid. She prefers to think of it as 'liquid gold'. One use for the system is to try to get a better idea of the microbes that make up healthy gut flora. Like faecal microbial transplants, her group's experiments begin with a sample of donor stool. But the set-up allows them to do new things with it. Samples from different donors can be mixed, for example, and the researchers can monitor which microbes persist when the system reaches a steady state. If one person's microbial ecosystem ends up dominating over someone else's donation, that is interesting to analyse.

The idea is to simplify possible therapeutic inoculations of gut microbes. At the same time, the approach avoids oversimplifying. It begins with what can be assumed to be viable population blends, produced by a real gut. 'In ecology, it is not necessarily all that straightforward to take microbes that have never played together and expect them to form an ecosystem,' according to Allen-Varcoe.

The chemostat has already yielded a carefully formulated brew that can be used as a substitute for a whole faecal transplant. A healthy donor's complete sample was reduced to 70 cultured bacterial strains, from which 33 were selected as good candidates for a simplified ecosystem.

The idea of using a known microbial mixture is not new. A ten-species blend was tried out on half a dozen *C. difficile* patients as long ago as 1989, and worked well, but the work was not followed up. The difference now, as Allen-Varcoe puts it, is 'we allowed nature to preselect our ecosystem resource'.

The 33 selected strains did form a stable community in the Robogut, and the same culture blend cured two trial patients with *C. difficile* infection who did not respond to other treatment. The clinical use of the new mix was then stopped by the Canadian authorities while they pondered how to regulate such a brew, and at the time of writing has yet to resume.

The unfamiliarity of messing with microbes, even if actual administration of faeces is not involved, will create headaches for medical bureaucrats for a long time to come. Does every new preparation have to be certified as fit for use in people? Who will do that? And is it likely that a complex mix of growing bacteria will not change composition? If the regulatory system set up to help ensure drugs or foodstuffs are safe is applied to preparations like this, and it is hard to see an alternative at the moment, the process of agreeing how to do it will move slowly, one feels.

Still, this tiny trial got heaps of publicity, partly because the lab came up with the name 'RePOOPulate' for the project. It was reported by the normally staid *Scientific American* under the immortal heading 'Scientists find they can make sh*t up'. And Allen-Varcoe sees it as proof of principle for a new approach to medical treatment she calls Microbial Ecosystem Therapeutics (MET).[14] Her vision is that a doctor will eventually have 'a panel of ecosystems' to choose from, and will select one that suits the patient. The work done so far is a small step in that direction, but does make it look like one possible future.

Our future microbiome

People are already trying other approaches to providing bacterial consortia instead of a full stool transplant. A report as I write features fifteen patients given a new product that includes fifteen bacterial strains – but delivers bacterial spores in capsule form rather than live cells.[15] And, although it is very early days, it does seem likely that we will in future be able to alter our various companion ecosystems to suit ourselves. Now we know so much more about their make-up, and about how important they can be, there is no shortage of people speculating about what form those alterations might take. So let's round off this chapter by letting speculation run free, with the caveat that realising most of these alterations is generally likely to be decades rather than years away.

I have no doubt that in the end people will want to make use of their microbiome. Some of the uses will be mundane – like preventing tooth decay (you can already try probiotic gum), dermatitis (probiotic ointment) or acne. Another that is already under test is a microbial transplant, from armpit to armpit. It is designed to help people with smelly armpits, by transferring skin bacteria from someone who finds their body odour easier to manage. One reason for persistent odour is that conventional antiperspirants are designed to mask the smell bacteria make, not eliminate it. In fact, Chris Callewaert of the University of Ghent finds that they tend to increase *Actinobacteria*, which are the main generators of body odour.[16] Calleweart, who also maintains the website drarmpit.com, is now experimenting with bacterial transfers to help reduce severe armpit odour, but has not published results yet.

Other, less mundane uses of modified microbiota are bound to come. We cannot say yet how exotic they might be, but we can look around for clues.

We already read stories in the media every day advising on the pros and cons of this or that diet or exercise regime. Like me, you probably try to make sense of the sum of these messages and follow some of the advice, or try to, or resolve to try soon. We have got the message, in other words, that we can maintain our bodies more or less well, and we would like to keep them healthy, or at least slow their deterioration. Some of the responses to that message support a global industry worth billions that sells vitamins, supplements, health foods, diet books and exercise apps.

Another branch of the self-maintenance industry caters to our wish to look attractive and fashionable (or, in my case, just vaguely presentable, really) and manufactures perfumes, hair-restorers, and cosmetics by the bucketload.

If you want to go further and make your body a work of art you can choose between a range of services from the local tattoo and piercings parlour to an upmarket cosmetic surgery clinic. Or you may want to explore drugs, old and new, that promise everything from altered states of consciousness to enhanced sexual performance to just staying awake longer so you can revise for that exam.

All in all, some of us, at least, are willing to apply *any* technology that becomes available to ourselves. Let's be up to date and call it body hacking.

As we get used to the idea that our bodies are superorganisms, the search for new ways of optimising or enhancing *all* the cells that are part of us will be on in earnest. The probiotics market already caters to people who want to get the best population mix in their gut.

As we adopt more refined approaches to managing the microbiome, one key question is how soon we turn to genetically manipulating denizens of the microbiome. We have had the technology to alter genes for decades now – it began in bacteria after all, and a flourishing biotech industry shows that it works. But suppose that the tailor-made bugs are destined to live on us?

In some ways, that seems OK. We are wary of applying GM to ourselves (eugenics! playing God!). Experiments on human cells need strong medical justification and dogged compliance with regulations.

The same restraints hardly apply to bacteria.* We know how to mess with their DNA, and we know we have innumerable bacteria in us. So it seems safe to predict that the first widespread applications of GM in humans won't be to our cells but to the other elements of our superorganism, our microbial fellow travellers.

There have already been efforts to develop altered microbes for uses outside the gut. It is possible, for example, to engineer *Streptococcus mutans* so that it still colonises the mouth,

---

* The first published report of reprogrammed bacteria being raised in the gut (in mice) appeared in mid-2014. The authors make it plain their near-term target is using them in humans. Kotula (2014)

specifically the surface of the teeth, but no longer produces lactic acid from sugar. Result: no tooth decay.

There's good evidence that this works. However, full-scale trials have been held up by the US Food and Drug Administration's decision to treat this as a full-scale experiment in genetic engineering. While it is not human genetic engineering, the result is intended to be *applied* to humans – and the climate for doing that has altered in recent years because of fears of bio-terrorism. When the first group to propose a serious trial of decay prevention using engineered bugs applied to the FDA for approval, they found their new bacterial strain classified with potential bioweapons.[17]

Some years on, the company the researchers began in the early 2000s still has the product on their website but there is no sign yet of it coming to market.[18] An alternative route, a mouthwash that contains a specific antimicrobial peptide so that it kills *S. mutans* without laying waste to the rest of the oral microbiome, might be a better bet.[19]

Meanwhile, the same company who are engineering *S. mutans* rather than trying to eliminate it are selling oral probiotics for humans and, if they wish, their pets. Health claims are largely ruled out by the regulations on marketing probiotics as food products, in the USA at any rate, so the pitch here is mainly cosmetic, even for pets. Sprinkle a patented blend of three microorganisms in powder form on your dog or cat food once a day and the beast will have fresher breath and whiter teeth.

As products like this become more familiar, might the barriers to incorporating engineered organisms be overcome, or will they stay stuck in the regulatory mud? One possibility is that the self-help motivation behind much consumption of health foods and probiotics could transfer to home-brew genetic manipulation, which is a now a relatively accessible technology. What we learnt to call 'genetic engineering' in the 1970s and 1980s is now being rebranded as 'synthetic biology'. The implication is that the production of new organisms that will realise human projects is now a matter of design. There is a grain of truth in this, for bacteria at any rate, with synthetic biologists' development of standardised 'bio-bricks'. These can be used as a communal kit of parts to construct microbes with DNA modifications tailored to a variety of purposes. This goes along with a culture that is in some ways more like that of computer hackers or open-source software developers than old-school molecular biologists (though they have always been good at sharing phages and strains of *E. coli*).

Here, two opposing trends collide, in ways that create potential for confusion. On one hand, there is a community of biohackers who are interested in using the tools of molecular biology, among other things, for their own projects. As the Wiki page of the biology enthusiasts who work out of the London Hack Space in Hackney (really!) says: 'The field and community is growing all the time, together with the ability of amateurs to do cool stuff.'

On the other hand, assembling the kit to manipulate DNA and using it to alter the genes of anything, even a few bacteria, might bring an anti-terrorist team in hazard suits to your door with battering rams and a warrant to arrest you and impound your home-brew lab.[20]

The twist in the tale could be that security concerns accelerate development in labs that are located openly on campus rather than in the back streets. Synthetic biologist Jeff Tabor works at Rice University in Texas on building strains of *E. coli* that one day might, say, detect metabolic changes in the gut, and respond with synthesis of compounds that help prevent obesity. A fantasy? The US Office of Naval Research appears not to think so, and recently gave him a half-million-dollar grant to develop the work.[21]

That old workhorse *E. coli* is the main target for synthetic biologists at the moment as they need a common property to work with and it is well understood. The work on human applications thus tends to focus on what a single bacterium might do rather than on, for instance, restructuring the population of the gut. It does generate some ambitious ideas, though. So far these mainly have potential health applications. John March at Cornell University envisages using *E. coli* to signal stem cells in the intestine to mature into insulin-secreting cells that could substitute for depleted islet cells in the pancreas.

The approach, he emphasises, is entirely general. It is worth quoting from his lab's web page to give an idea of the scope of synthetic biologists' ambitions, which look toward 'the development of modular expression cassettes that will re-configure target organisms for safe and effective therapeutic synthesis within a mammalian host'.

They go on: 'We are engineering enteric bacteria into effective *in vivo* cellular factories, responding to a specific molecular imbalance by synthesizing an appropriate corrective therapeutic. Cellular therapies share a need for accurate detection of target molecule levels, benign coexistence within the host, and a sufficient level of tunable gene expression.' In other words they envisage tweaking bacteria so that they can sense when something is wrong inside your body, make the medicine indicated, and deliver it, all in one self-sustaining cellular package. That is quite a programme, guys.

The ambition could be broader still if we imagine stretching the definition of therapeutic. Consider, for example, the reports of the odd individual whose gut microbiome is colonised with brewer's yeast. The result: 'auto-brewery syndrome', in which eating sugary or starchy foods leads to intoxication with alcohol made during digestion. The case reports here concern people with an unusual overgrowth of yeast after heavy doses of antibiotics or because their immune system was suppressed. However, we already know the normal microbiome is important for processing many drugs. What if it can be modified to produce them as well, on demand? Perhaps adventurous early adopters will dose themselves with a bacterial cocktail before departing for a party. Once there they will simply nibble on the foods their new microbes require to get them high. People already self-experiment enthusiastically to attain a legal high. Suppose a combination of a legal prebiotic and a microbe that metabolises it into something more pharmacologically interesting were on offer? There is much amusement to be had from contemplating the authorities who would try to police such things, never mind the other effects.

Working from the outside

For those who draw back from allowing modified bacteria to join the microbial party in their intestines, there are other approaches. The total faecal transplant is not going to be top of the chart, but there are other transfers that are less difficult to implement. Someone getting persistent ear infections might try a little earwax from a more fortunate friend. Perhaps women with vaginal infections will meet to acquire vaginal swabs from an infection-free helper. A baby shower could be transformed into a microbiome party, where the family makes sure everyone gets to handle the baby as much as possible. You are especially welcome if you bring your dog.

The next step is products that supply the right bacteria to anoint yourself with. A spray-on for skin care, intended as a substitute for soap and deodorant, is already under development and has been tried by a few intrepid journalists.[22] The company bringing it to market, AOBiome, says that its patented ammonia-oxidising bacteria 'have been shown in a cosmetic clinical trial to improve the appearance and feel of people's skin'. It quotes a price of $99 for a month's supply, but has yet to scale up production so has a near three-month waiting time for orders in the US.

Further ahead, imagine a future in which we spray our homes with bacteria instead of disinfectant, bring home-fermented foods into hospital to strengthen a loved one's recovery, or dab a cunning mix of bacteria behind each ear to boost our pheromones before a date. As the vision of a microbiome party suggests, future uses of our microbiome will be shaped by social as much as scientific or technical developments. Commercial influences will be in the mix, too. That bacterial skin-care spray will be advertised as an aid to family health. Items you touch a lot, like the steering wheel in your car, will have probiotics that help skin care added as a special manufacturer's upgrade.[23] Your mobile phone might come designed to harbour desirable microbes, suggests built environment specialist Jessica Green of the University of Oregon.[24] How will we react to all this?

That will depend partly on other shifts in attitude. There are avant-garde architects and designers keen to experiment with bio-engineered materials for buildings, on a scale from domestic experimentation to large-scale urban planning.[25]

Thinkers like Rachel Armstrong speculate about a 'living architecture' in which structures are grown rather than built. There are already experiments with self-healing concrete, made by including bacteria in the mix that secrete limestone to fill cracks. The advantages – ecological, aesthetic, perhaps (who knows?) immunological – might outweigh any distaste for entering buildings that are themselves living. Will a spreading awareness of our own continual microbial interactions with our surroundings make us more relaxed about such things? Or will people newly sensitised to the importance of their microbiome grow more wary about encounters with new, bioengineered materials, *especially* if they are still alive? We can only guess.

There is conflict in store as changes in attitudes to microbes play out. There is no reason to think that everyone will simultaneously shift their opinions and decide to love germs, or will agree which ones to encourage. Perhaps early signs of the kind of discussion we will get more accustomed to can be read in recent disputes over a microbial product lots of people love – cheese. Specifically, there are arguments between regulatory authorities, who try to eliminate unpasteurised cheese because it can harbour pathogens, and cheese connoisseurs and cheesemakers who want to use raw milk to produce more delicious results. As one anthropologist puts it, in the US, 'while the Food and Drug Administration (FDA) views raw-milk cheese as a potential biohazard, riddled with bad bugs, aficionados see it as the reverse: as a traditional food processed for safety by the action of "good" microorganisms—bacteria, yeast, mold—on proteins found in milk.'[26] Some of the raw-milk cheese enthusiasts support a version of the hygiene hypothesis and believe that eating a richly microbially-endowed cheese is healthy because it boosts the diversity of intestinal flora.

As we begin to see efforts to fashion a bioengineered domestic or urban landscape, familiarity with the microbiome will affect our reaction to other products, old and new. It will probably be encouraged first of all by including our microbiota with other aspects of our biology and physiology that we are invited to monitor daily. Consider, for instance, the 'microbial home' dreamed up as part of a futuristic design study by the Philips company in 2011. The Philips futurists imagine a morning routine that includes taking note of a battery of new devices in the bathroom. There might be sensors on the mirror that analyse your breath for telltale chemicals, the results to be interpreted in conjunction with microbial samples recovered from your toothbrush. The lavatory would automatically analyse a sample of your urine and faeces, and the shower keep tabs on skin microbiota. All this is not Orwellian, but takes place in a world where 'It is part of the daily ritual, a self-awareness tool, part of the grooming and cleansing process'.[27]

What about actual grooming, then? Augmenting gut bacteria by sprinkling a probiotic powder on your food or swallowing a capsule is an easy sell. Taking up microbial gardening in our other ecosystems is likely to need something more like cosmetic or deodorant products we already use – perhaps rubbing in a lotion or a paste, using a spray or a roll-on applicator. Will that seem OK with bacterial preparations? Some entrepreneurs are already hopeful. Gilad Gome, founder of a California start-up called Personal Probiotics, begins prosaically but then gets more fanciful in an interview on the website Motherboard. vice.com. He suggests that a woman could protect herself from urinary tract infections and pathogens by taking a probiotic, and that the bacterial strains used might also be modified so that 'if she wants she can hack into her microbiome and make her vagina smell like roses and taste like diet cola'.[28] Whether or not this is a serious proposition, it is an effective way to get attention for a new company.

Synthetic biologists have worked with artists to devise projects that help us imagine how we might come to use such products. Some of the items they offer us as objects to think with are as yet imaginary. Would you use a personal skin care product, for example, that delivered living microbes in a small container where they feed off cotton balls? The microbes, the artists suggest, 'could also produce fragrances, soap, and oil molecules, vitamins and signalling molecules in a combination that is most appropriate for the individual's skin type.'

Another effort from the same project, though, was a new application of the human microbiota made real.

Mapping disgust

How do you like your cheese? Some of us like it strong-smelling as well as strong-tasting, each enhancing the other. But if, like me, you are one of them, chances are you don't much care for cheesy feet.

Odd, that, as the bacteria that produce the smell are almost indistinguishable. Limburger cheese, for example, is made using *Brevibacterium linens*, a near-identical sibling of *Brevibacterium epidermis* that lives, if we let it, between our toes. Put a piece of Limburger cheese outdoors and mosquitoes will gather, under the mistaken impression they are homing in on a good feed from a foot.

Lots of other kinds of cheese get their distinct aroma from bacteria that can live on us. We will never know, but it seems likely this is how some of them got into cheese vats in the first place. Their presence in our diet shows how the way we react to smells and tastes associated with decay depends heavily on the context.

The same art–science collaboration mentioned above explored the shifting boundary between 'yum' and 'yeeurgh'. One of the teams on the project called Synthetic Aesthetics, Christina Agapakis and Sissel Tolaas, used microbes from project participants' skin, recovered on swabs from armpits, hands, toes and noses, to ferment milk.[29] Reassuringly, perhaps, it was 'organic pasteurised whole milk'. After an overnight incubation, they found cheese-in-the-making, and did what regular cheese-makers do, straining curds from whey. The result was a collection of eight cheeses, 'not the aged masterpieces of artisan cheese-makers, but microbial sketches, capturing some of the ecological diversity of different bodies and different body parts.' They produced, they report, 'a wide range of smells'.

It would be unwise to eat any of them, as their initial bacterial load was undefined – although each one was analysed in detail later on. But the reaction on passing them round a lecture theatre to sniff was still revealing. As one of the instigators of the project, Daisy Ginsberg, recalls, 'The audience passed them along, tentatively smelling each cheese, some trying to distinguish themselves with a measured air of objectivity by comparing smelling notes with their neighbours. For others, the small, white, moist cheeses were just too much, transgressing accepted boundaries of decency.' She reports, tongue-in-cheek, that 'Daisy Armpit' was judged the best-smelling preparation, with 'a fresh, yogurt-like aroma'.

In contrast with the wary audience reaction in the lecture theatre (ironically – or not – at the high temple of geekdom, MIT, in Boston), the staff in a Berlin cheese shop who also smelt the samples were 'marvellously unperturbed' when told subsequently how they had been made. Christina Agapakis also found that people were less likely to be repulsed by their 'own' cheese than by one made with bacteria taken from someone else.

So if you become more intimately acquainted with micro-organisms and how they do their work, will disgust moderate? At the moment, as the science fiction writer Bruce Sterling put it, 'We lack a positive way to describe a joyful and life-enhancing infestation of our flesh by tiny microbes.' But he, too, foresees a cultural shift as knowledge of the microbiome spreads. In the future, he predicts, You'll be 'into germs because germs are into you'.[30]

How fast will this effect spread as we become educated about our microbiomes? Well, we shall find out. The future of the microbiome will surely be different, as we apply our emerging understanding of how important it is, and how it works. But for now, let me go back to the meaning of the microbiome in the present.

12 | I, superorganism?

My life, like yours, is a palimpsest. I am biological, to start with: a creature. I have creaturely needs. Yet the way I experience them is through overlapping layers of cultural, social and technological mesh. And in the life I live – probably yours too, unless you are an athlete in training or you have a serious illness – the social and cultural usually seem to me to matter most.

So what difference does it make to get better informed about the biological part of me that happens to be composed of microbes? I suggested that contemporary research is characteristically modern in the sense that it treats the microbiome as an enormous store of information. I can confirm there is a *lot* of information, encountered while exploring what the research has been investigating and how it gets done – certainly more than I could fit in here. But then nowadays there is a great deal of information about practically everything. If we now stand back and consider what we have learnt, we need to decide how important microbial info is against that background, as well as which parts to pay attention to.

In some ways, the microbiome is overwhelming – so many, many, tiny cells. Almost whichever way you look at it, the collective they form with us is unspeakably complex. Add the rapid rate of publication and it would be idle to pretend that I have all the answers at this point, or even a decent proportion of them. But I will try to relate what I have learnt, in three ways. One thing you discover when reading reviews and listening to scientists is what the interesting questions are, so we will look at some of these shortly. I also want to return to the idea I started out with, and turn it into a final Big Question: am I a superorganism, and what do I think about that? First, though, let me give you some advice. By now, you can decide for yourself whether it counts as informed advice ...

Managing a modern microbiome

This is not a self-help book. At least I didn't think it was going to be. But I have learnt some things along the way that made me think understanding might lead to action. If you've been skipping, here are the things that currently seem worth considering, as far as I can judge the state of the science.

On diet, everything we know about the gut microbiome suggests there is wisdom in the food writer Michael Pollan's brilliant formulation summing up all the best advice on what to put on your plate: 'eat food, not too much, mostly plants'.[1]

Many would add one more thing: make sure some of the food is fermented. I still doubt that is strictly necessary. It might be better to make sure you eat as many different kinds of vegetable fibre as possible – as a prebiotic – and promote an equal opportunity gut that way. Still, fermented foods may do some extra good, and probably no harm, so throw some in as you see fit. Some of them are quite nice.

The ones you choose will depend on personal preference and ease of access. Yoghurt in the morning suits me – Metchnikoff would approve. I have a deep, discreditable prejudice that sauerkraut is peasant food and I cannot abide kimchi, but am very willing to eat unlimited quantities of blue cheese, preferably accompanied by plentiful fermented grape products. These, however, are merely, ahem, cultural preferences.

I've pretty well always been healthy, so far. If I'd ever had Crohn's disease, irritable bowel syndrome, or gluten intolerance, I'd take more detailed advice on probiotics. The best way to find out what in the way of probiotics might help, though, is self-experimentation. Individual responses vary widely. So do the contents of probiotic products. It's best to view manufacturers' claims with a raised eyebrow, take careful note of exactly what you try, and when, and of any changes in your condition. If it changes for the better, the new ingredients in your diet still may not be the reason, but at that point why would you care? Everyone has the right to act on their own anecdotal evidence. The mistake is to expect much guidance from other people's.

If you have the cash and like looking at graphs, it is possible now to track changes in your personal microbiome as you experiment with your diet using one of the crowd-funded mass research projects now under way. The results will almost certainly be interesting. They will also be a bugger to interpret. Your call.

Probiotics are one thing. If you do have a chronic bowel problem, you may be tempted by other measures. My advice, for what it's worth: don't even think about doing a faecal transplant at home. This is a desperate remedy, for life-threatening afflictions like chronic *C. difficile* infection – the only condition so far where there is convincing evidence of benefit. It calls for medical supervision, and donor screening. With luck, by the time you really must have it, the researchers and regulators will have got their act together and found ways to develop standardised culture mixes that do the job without resorting to such a haphazard, uncontrolled and, let us admit, disgusting for all involved, procedure.

Now babies. There is reason to think caesarean delivery has costs that were not recognised before we knew how it alters the early microbiota. The easiest, safest and most obvious way to guard against them is the vaginal swab. It may not be the first thing on mother's mind after abdominal surgery, and someone may need to speak persuasively to a midwife, nurse or doctor, but it seems worth doing. Maybe it does not have to be done straightaway and you can arrange a ritual anointing when baby comes home?

It is fantastically annoying and demoralising to be told that formula milk is not great when you have a massive mammary abscess,* a baby who won't latch on or a problem with milk supply but, from the microbial point of view, breast is best. In time, there will be ways round that, which supplement artificial milk with tried-and-tested bacterial cultures and prebiotics – but we're not quite there yet.

And once your beautiful baby has acquired a well-balanced microbiome, the fewer times you assault it with broad-spectrum antibiotics, the better. It would be crazy to abjure these wonderful drugs when your child – or anyone else – has a life-threatening infection. If it is lower down the scale, consider alternatives, even if the poor wee mite is getting some discomfort. It will pass. Trading off pain relief now for problems avoided later on, possibly much later, is hard, but it is what

---

* I speak from experience here – one involving fever and delirium, followed by emergency surgery for mother. We all survived but it was pretty scary at the time.

most of us teach our children so maybe we can try to practise what we preach a little more.

The medical profession needs to reconsider its attitude to antibiotics, too – a process that is already under way – and to make sure this extends to adults, and especially the elderly (more and more of them) as well as children.

Off prescription, mouthwash and antibacterial soaps will probably do you more harm than good; swabbing your skin with astringent cleansers if you have acne may make things worse (it certainly never did me much good, and I had it pretty badly). Treat advertising that continues to exhort us to deodorise, sanitise and disinfect every bodily and household surface with scorn. Scientifically informed scorn.

Oh, and allowing farmers to use sub-clinical antibiotic regimens to fatten livestock is completely daft. It must be banned where it is still permitted, and bans actually enforced where they are already in place.

Finally, as we take a new vow not to wage war on the microbiome, that does not mean losing all care for hygiene. There *are* still pathogens out there. You do not want to eliminate bacteria from your skin, for instance, as that just makes colonisation by less desirable species more likely. But soap and water will not do that – they seem to be better at removing microbes that have not yet got established. So yes, you should still wash your hands.

I have questions

For me, the addition of all those trillions of microbes to the system that both constitutes and sustains me adds immeasurably to the personal fascination of biology. Much of it remains mysterious, not in a 'there are forces at work that we cannot comprehend' way, but more in the sense of, 'gosh there is a *lot* here that still needs working out'.

Which is to say that after reading as many scientific papers on the subject as I could lay my hands on I still have questions. That is not surprising. My superorganism is a young thing – not, alas, because I am, but scientifically speaking. We have known for a while that we contain microorganisms, but the current techniques for investigating the microbiome have not been around for long. After summarising what we now know about all these fellow travellers and what they do for us, I am keenly aware that this is a developing field. Researchers know more than they did last year. Next year they will know a lot more. At least, they will have a lot more information, although clearer understanding, as ever, may take a little longer to achieve.

I hope you agree at this point that we already have enough new findings to prompt an extended look at the microbiome. But there are plenty of open questions, too. Here are some of the most important ones that are likely to find new answers over the next few years.[2]

Will there be enterotypes, and what will they tell us?

This query about whether there are a few common types of gut bacterial mix that are stable and do all the normal favours we hope for from our microbes in digestion, provision of vitamins and prevention of disease, stands for a larger set of questions. How much individual variation is there in the gut microbiome, and how might it relate to groups of people – whether defined by particular characteristics of the microbiota, or by more traditional classifiers such as gender, age, class, ethnic group or culinary culture? If we can unravel that tangle, it will be a useful step on the way to knowing what different kinds of healthy microbiome there are, and what an unhealthy one looks like.

Will we be offered 'metagenotyping', and what will it tell us?

You can already get a rough population snapshot of your microbiome, and it can be repeated over time if you want. Doing more intensive DNA analysis is certainly possible in principle, but not yet a sensible thing to do outside funded research. But in the age of ever-cheaper genomics, perhaps a full gene readout of the microbiome will be on offer to everyone. What advice we will need to interpret it, and perhaps use it to help with any personal decisions – dietary or otherwise – is still pretty unclear.

How should we handle antibiotics?

There is good reason to get much choosier about prescribing antibiotics, and which ones to use. Does their effect on our normal microbes offer a stronger reason for change than the one that is already clear from the ever-growing threat of antibiotic resistance? It may well do, but we need to know a lot more to figure out how to get the benefits of antibiotics without the drawbacks. We might, for instance, move to sampling gut microbiota before giving the drugs, to assess whether the patient has a population mix that is more or less likely to recover after being zapped. But that will only be possible if there is no medical emergency. Elderly patients might be screened for *C. difficile*, which can be present in low numbers in healthy people. Even there, people who develop problems with the bacterium later on might do so because antibiotics allowed a pre-existing population to grow much larger, or because they were colonised by a different strain after treatment.[3]

When will there be treatments based on new microbiome studies, and what will they be like?

There will obviously be medical developments, both for treatment and prevention, from all this work on the microbiome. But the road to what people in the business call 'translation' from the lab to the clinic, or from bench to bedside, may be long. Most results still come from mice. Many procedures and data-processing protocols are far from standardised, which makes it harder to bundle studies together to firm up conclusions. We can expect 'swallow it and see' treatments to appear, which will probably be hard to evaluate. More precise efforts will probably still require detailed, mechanistic, preferably molecular understanding of what is going on. In most cases, higher-level microbiome studies have only raised suggestions about that, which remain to be tested. Perhaps early trials will involve microbial-derived metabolites, or cell-surface chemicals separated from their home species.

Will there be more surprises about what microbes can do for us?

Almost certainly yes. By definition we do not know what they will be, but a good guess of one area they will appear is epigenetics, the newly fashionable study of how our own genes are modified in subtle ways after we are conceived. This is itself a rapidly emerging field, full of surprises, concerning the way the basic DNA code in the genes is annotated by adding or subtracting accessory chemical groups to the bases in a genetic sequence. These changes, most often involving addition of methyl (CH_3) groups at key sites, are a little like a collection of software preference settings in a computer program. They determine which functions and sub-functions get used, and when, and who has access to particular routines. They can affect whether particular genes or collections of genes can be expressed at all. Epigenetic markers are put in place as a result of experiences the organism has, which translate into cellular responses that result in alterations to the genome. There are already hints that inhabitants of the microbiome can have epigenetic effects, in the immune system and probably elsewhere.

How long will it take to piece together a detailed picture of the microbiome's influences on our immune system?

The biggest thing the microbiome has done so far, to my mind, is induce such a large shift in thinking about the origin and function of the immune system. I have tried to capture something of what it means. But there are loads of unanswered questions about the nuts and bolts. How do microbes in the gut affect differentiation of immune cells, especially T cells? Why do some species do this, but not others? Will we be able to show direct effects on allergy or autoimmunity, or will they just remain associations that tantalise?

And finally ... what else do we need to know?

That question is always in the air when the science is interesting. In the life sciences at the moment, the general answer that is usually given is that we need to know how to develop 'systems biology'. What's that? It often seems to be discussed with lots of hand-waving and a vague impression that it will be a miracle ingredient or universal solvent that will allow us to tackle all our unsolved problems. It really means looking at how all the various bits and pieces of a complex organism work together – a kind of super-physiology, integrated at every level. Reductionism will help to pick apart a particular system, but there always comes a point where you have to try and put Humpty Dumpty back together again.

That understandable general goal has spurred some hugely impressive efforts to take the measure of defined bits of life. The simplest of cells, *Mycoplasma genitalium*, has been captured in a computer model that incorporates every bacterial gene and its product and follows their interactions through an entire cell cycle, from a newly duplicated cell to its next round of division.[4] That was a massive undertaking, and filled in lots of data gaps using work on other species like *E. coli*. It is a first draft, for one cell of one species, and the results for things like levels of particular chemicals or rates of synthesis only approximate those found in actual cells, but it is a landmark effort to simulate a whole cell. At the other end of the scale, there is a pretty convincing computer simulation of the beating heart, its cells coordinated by electrical impulses, built from the lifelong study of the British physiologist Denis Noble.[5]

But with the advent of enormous heaps of data about, well, practically everything, a convincing vision of systems biology for a superorganism is still hard to find. It seems to be talked about more often now, while managing simultaneously to recede from view. There are many models of interactions between two or three bacterial species, and their metabolic competition or cooperation. Incorporating other aspects of cellular life, like signalling, still less interaction with complex hosts, is some way off.[6]

At the moment, there are useful proofs of principle that show, for example, some of the ways our intestine might select the right bacterial species. They convince by simplifying drastically. Jonas Schluter and Kevin Foster of Oxford University are trying to explain how a community of bacteria that generates benefits for the host is maintained. In theory, any bacteria that simply gobble up all the nutrients they can get at without providing anything for any other cells to use will have a competitive advantage. Such freeloaders will grow faster, thus coming to dominate the population, posing a problem for theorists who want to account for how co-operation can be maintained. The Oxford team showed in a mathematical model that the way bacteria concentrate near the intestinal epithelium means that small selective influences from host epithelial cells – whether release of key nutrients or of antibacterial compounds that affect only some species – are amplified, and can counteract the theoretical advantage of bacterial cheaters. Their model, though, contains just two species, not thousands, and does not represent any of the details of bacterial metabolism, so its insights are just a beginning.[7]

One way of imagining how an ultimate systems biology in a complex organism could work is that it will capture how all four grand information and control systems I have highlighted – genetic, hormonal, neuronal, immunological – interact, and thus regulate metabolism, growth and cellular and tissue development. No small task. The new-found appreciation of the microbiome adds to the conceptual burden here, as we've already seen. Now we have to fathom how the microbiota affect, and are affected by all these systems, too. Only then might we see a true systems biology of a superorganism.

There are plenty of comments in review papers that point in this direction. But they are prospective, and stop short of showing the way in detail. One says we need 'a more comprehensive map of genetic factors involved in host-microbial and intermicrobial crosstalk' and of 'interdependencies of microbial species and the network architectures of intestinal colonization'

– that is a nod to systems biology. The remaining question is then whether it will emerge from a 'more comprehensive map', or whether we will see new concepts applied to understand the total system. I'd love to know.

Mirror, mirror

Modern cosmology has shown that all the elements in the chemists' periodic table beyond helium are made in stars, or – for yet heavier ones – supernova explosions. Scattered through space, they allow complex chemistry, hence life, to happen when they coalesce to form planets. Truly, as Joni Mitchell sang in 'Woodstock', we are stardust.

Martin Rees, the UK's Astronomer Royal and a science populariser with a nice sense of mischief, has another take on this. We are, he likes to tell people, nuclear slag.

His point, I think, is not that we are worthless rubbish rather than a wondrous product of a process that suggests we are somehow meant to be here. Rather, he is saying that the facts do not speak for themselves. Scientists describe what is the case, as near as they can figure out. Then we attach what meanings to it we can – and we can choose, or at least have some say in which interpretations are most persuasive.

That is what happens when a new set of discoveries kindle wide interest. The flood of new findings about the life that lives on us shows why. There is no one way to turn them from the technical language of papers in the journals into stories we can all understand. At least to start with, the story is up for grabs.

You can see that in the sudden outgrowth of new metaphors when early reports of the new microbiome studies appeared. We heard about 'aliens inside us' and a 'bacterial nation', 'a legion of little helpers', our 'bacterial buddies', microbial signatures and fingerprints, and a new extended self. Scientists were the source of these conceits as often as journalists.[8]

So as I come to the end of this particular exploration of my extended self, what's my story? Well, let me look at myself again, knowing what I know now about the microscopic interactions that are part of me.

* * *

Here I am, in front of the mirror again. What has changed? This old body looks the same. Am I thinking about it differently? Yes.

If I try to be more precise about the difference, it comes down to this, I think. In a way, it seems more alive. That sounds odd, even to me. I certainly felt alive before. I'll try and explain, to myself really, what I mean.

Before, I suppose I thought of myself as basically one big thing – an adult human, a pile of flesh on the way to becoming a carcass but doing alright so far. A late baby-boomer, and a bookish one, a person who has had good luck and good health, and a man; I rarely thought about the body as such. My early scientific education in biochemistry gave me a lasting appreciation for the intricate beauty of its molecular machinery. I saw it, though, mainly as machinery. It was a living assembly that needed to be maintained, and allowed the right inputs and outputs. What happened to them was really all about the chemistry. The inputs interest me, because food is nice. The outputs, not so much – aside from those moments one doesn't talk about when there is an unsought flash of infantile glee at producing an especially nicely formed turd, that lasts a second or two before it is flushed away. What happened in between was the breaking down of food into small molecular components, as a prelude to its rebuilding into molecules of me, a process summarised in the vast charts that adorn biochemists' office walls labelled 'intermediary metabolism'.

In the 1970s, some news from biology jolted this view. The compelling evidence that those energy-generating organelles in the cells of eukaryotic organisms, the mitochondria, are descended from ancient bacteria that have made a new way of life inside other cells pointed to a new picture of multicellular organisms, including us. We were still an immense cooperative assembly of cells, but each of them harboured other entities, somewhat disabled mini-cells, with their own small set of genes and distinctive membranes, and protein-synthesising machinery. Along with the great essayist Lewis Thomas I marvelled that I seemed to be 'a very large, motile colony of respiring bacteria, operating a complex system of nuclei, microtubules, and neurons for the pleasure and sustenance of their families'.

This is a startling thought. And yet somehow easily forgotten. These myriad endosymbionts have long since lost the capacity for independent existence. Our association is so intimate and ancient that my intuitive sense is still that they are part of me, not some separate, cohabiting colony. Elucidating the origin of mitochondria – and the similar story of the chloroplasts in the leaves of plants – induces a sense of wonder about evolution. But knowing that these invisible passengers are there remains a bit abstract.

With the microbiota it is different. I still can't see them, individually. But I know they are constantly arriving, reproducing, and departing. I feel more *inhabited*. There is almost a temptation to agree that all those microbes in me should be seen as 'trapped in huge colonies, locked inside highly intelligent beings, moulded by the outside world, communicating with it by complex processes, through which, blindly, as if by magic, function emerges'.[9] Almost.

I am certainly newly struck by the *size* of the colonies. I find it hard to visualise what the kilo of microbial gloop in my colon looks like, but knowing those (usually well-formed) turds are half microbial by weight seems impressive, somehow. If I now think of myself in part as a bioreactor – and I do – I'm a pretty productive one.

The quantity alone makes me think that the microbiome must have something to do with my well-being, or lack of it. And all that researchers have broadcast about the interactions between my tissues and the trillions of smaller cells that live among them does change how I think about 'me'.

For starters, I am no longer mechanical, or chemical, even though I know all the interactions that are going on are still governed by molecules. The fact that I provide a habitat for a large array of other species, whose populations rise and fall as they compete for resources, makes my interior life seem more *eventful*, in a very particular way.

It certainly makes me feel I should pay more attention to the inputs. I am not just taking on fuel, or even raw material and micronutrients, so that automated miracles of chemical breakdown and biosynthesis can build me new cells (though I am still doing that). I am eating to feed my microbiota – and sometimes adding to them as I eat. That feels different to me. The peristaltic gurgle I hear as I lie in bed after a good dinner is no longer just the muscular action of the intestine ensuring the passage of chewed food. It is the agitator for my main bioreactor.

That is not all it signifies. It is also a noise made by an ecosystem, one of several that I now have to reckon with. An ecosystem is different from an organism. It used to be something I thought of as the kind of thing that was external to me. An interior ecosystem is a different proposition. Perhaps it regulates itself readily enough without any care or attention from me. But I am inclined to give it at least some thought as I eat, drink, exercise or medicate myself.

But I do think, in the end, that 'bioreactor' or 'ecosystem' are not the most interesting ways to think about my extended microbial ensemble. The microbes and I together are, some maintain, a new kind of biological entity that deserves a technical name – I am a *holobiont*. I don't think that's going to catch on. No, I conclude that I really am a superorganism.

That is a term that has come up before in quite a few biological contexts, some of them inspired by findings in microbiology. Different species or strains of bacteria that share a pool of genes that can be drawn on communally can be considered as a superorganism, for instance. It has even been suggested that genetic exchange at all levels of life is so common that you can treat a global community of genomes as a kind of superorganism (one not to be confused with the planetary organism known as Gaia, incidentally).[10]

Neither of those quite captures the sense of superorganism I am now embracing. This one is a blend of one, highly organised cellular community – eukaryotic, multicellular me, I suppose – with an unimaginably large number of other, flightier living entities that choose to cohabit to maintain something larger but harder to define precisely.

Day to day, how does becoming conscious of being a superorganism of this kind make a difference? I am certainly not super in the sense of superior. I can assume my new status as superorganism only by taking on board that every other multicellular creature has a superorganismic side, too. But it does trigger wonderment at the way life has evolved to integrate large and small in the most intricate, intimate ways. The 'I' that is a superorganism has a fuzzier boundary than this body would have without its microbial load, and its make-up is probably more serendipitous, and less stable than I am used to. It is in some ways an elusive entity. It shimmers, seeming to afford a clear view, then slipping imperceptibly into something slightly different. But it is also more connected with the rest of the world, especially the biological world, than I ever appreciated. And so are all the other creatures on the planet.

I am struck also by the way in which a superorganism of my type is rather different from an organism. An organism is a thing that can be defined, biologically, reasonably clearly. It has a unified aspect, and you can decide whether something is part of it or can be regarded as separate. If it has parts, the whole is greater than the sum of those parts, and they are all coordinated by interactions that keep that whole in place. Departures from that usually signify illness. At cellular level, for example, a tumour can still be regarded as part of me – we share a genome, at least in the early stages of its growth. But I am pretty clear that our shared interests have diverged. If I can eliminate the tumour, and it is of the kind that leaves little visible alteration to my body, I am more likely to feel that my 'self' has been restored than diminished.

Some of this sense of a governing whole is carried over into yet another use of 'superorganism', to refer to social insects (and the odd mammal – hello, meerkats) that are genetically linked and function as an entity. They can, from some points of view, be treated as a single organism, and the parts as separable but not separately viable. They also have a definable make-up. We can tell which bee, for example, comes from which hive.

The social organism now coming into view as we reassemble me and my microbiome is not really like that. The microbes that are included are not tightly defined, or fixed – their precise population mix is highly variable, depends to some extent on random events, and shifts over time. It is a loose collaboration, with partners that change.[11] Many of the species or collections of species it contains would be perfectly happy to live elsewhere. They may look as if they are serving some larger purpose when they are in my gut, for example, but the bacteria there are growing because it suits them to live that way. The whole arrangement is highly contingent, and its stability can be compromised at any time. When I die, some of the bacteria I have been feeding will, with splendid microbial indifference, turn to eating me. We are connected, but our fates are not by any means indissolubly linked.[12]

The other big change in my outlook on myself is the way I am learning to think of those connections being mediated by the immune system. I realise that it has puzzled me for a long time. The military-cum-surveillance view of how it all works never really convinced me. The ideas now being developed by scientists looking seriously at how we live with such a complex microbiome turn out to be the explanation I was waiting for.

I now find myself thinking of the immune system, other cells and tissues, the microbiome, even the brain, as part of a continual conversation. Mostly it is an untroubled background murmur, cellular cocktail party gossip. Occasionally the tempo of talk steps up, and when the party gets really agitated there are arguments, and worse. Sometimes, there are terrible misunderstandings. The whole thing seems more nuanced, and richer than earlier versions. Conversations cover a range. Most just go something like: 'You alright?' 'Yeh, alright.' 'OK, then.' Sometimes they are more along the lines of, 'I think you might want to have a closer look at this', escalating to 'We could be in for trouble, here – send help, please', or perhaps even to a Dalek-like devastation: 'Emergency! Exterminate!' But more often everyone concentrates on being terribly polite because they want to avoid that.

This is just my fancy, and I do not know if I can influence the conversation. The language is not one I speak, and is in any case largely inaudible. But it does seem important to acknowledge that it is going on.

It is another way of being connected, and goes along with the other ties I now sense more acutely with microbial life – in my cells (as mitochondria), in my genes (as evolutionary survivors) and in my gut (as auxiliary members of a larger cellular community).

Those connections may be with forms of life that are too small to see, but I am delighted to be here at a time when science is making them visible in such arresting ways. So, comradely greetings to all of us superorganisms. I'm very glad to meet you.

* * *

There is one final way in which I choose to view my microbiome as making me more connected to the rest of life. I can put it grandly when the mood takes me. The sense of superorganism as a loosely woven mesh that unites all of life on Earth through microbial and viral exchange, and on occasion genetic exchange as well, appeals. I once wrote a little book about Jim Lovelock and his Gaia hypothesis,[13] and while I don't buy the idea of Gaia as some masterpiece of planetary self-regulation, I am keenly aware of the differences between this planet that is alive and those that are not. It is good to be part of it.

But this awareness also offers a nice, local simplification of the knotty problem of what the practical implications might be of cultivating a microbiome that is a small corner of this mesh. I think again of Graham Rook's take on the hygiene hypothesis, his idea that we need to stay acquainted with our 'old friends'. Green spaces are a good way to do that, he urges. Do I need to get up close to leaves and soil, birds or insects, then? Not necessarily.

Microbes, after all, are everywhere. Consider, says Rook, the populations of the air. Bacteria found in soil and water are also abundant in the atmosphere. They are found in the soil and air of cities, too. One early study in the US found 1,800 species in urban air.[14] I cannot forget that the soil in Central Park in New York, admittedly a large park, yielded an astonishing 170,000 microbial species in a recent survey.[15] I remember that above shrubs and grassland there can be getting on for a million organisms in every cubic metre of air. An average person inhales eight cubic metres a day. So, just maybe, I can forget about prebiotics, probiotics, or whether to get a dog in the interests of microbial diversity. All I need do is visit the park round the corner from my house, linger a while … and breathe.

Acknowledgements

Lots of people helped by mentioning things to read, watch or listen to. Thanks to all. Even more gratitude to those who read all or part of the drafts, and commented on work in progress so that I could try harder to make it more coherent, more readable, and ideally both. Where I am still off target, it is not their fault.

In no particular order, thanks for reader feedback to Margaret Boushell, Corra Boushell, Nick Lane, Nicholas Waller, Jack Stilgoe, Peter Washer, Peter D. Smith, Tania Hershman, Liz Kalaugher, Andy Extance, Kendra Brown, Tim Hayes, David Cameron (no, not that one), Margaret McFall-Ngai, and Felicity Mellor. And thanks for the same to a whole bunch of Turneys, who had less chance to decline: Eleanor, Catherine, Danielle, Mike and David.

The team at Icon who turned the draft into an actual book also encouraged various improvements. Thanks for that and more to Duncan Heath, Rob Sharman, Andrew Furlow and all.

Thanks, as usual, to my agent Louise Greenberg, and to the authors of the essential software that helps me manage this stuff – Scrivener, Evernote, Dropbox and, used heavily towards the final stages, desktop search on all the household Macs.

Also, because you do not usually get to acknowledge helpers who number in the trillions, thanks to all my invisible microbial collaborators, who kept this superorganism going long enough to write another book.

And thanks to Danielle, for everything else.

Notes

Chapter 1

1. Kroes (1999)
2. OGOD: a term I take from medical sociologist Peter Conrad, who suggests that germ theory and gene theory follow the same template. See Conrad (1999).
3. The story is well told in John Waller's *The Discovery of the Germ* (2002).
4. See Fowler (1986)
5. Nicholas Bakalar's *Where the Germs Are* (2003) is a good, balanced overview of good and bad microbes, though still emphasising the latter.
6. As noted by Jessica Snyder Sachs in *Good Germs, Bad Germs* (2007), p.29.

Chapter 2

1. The late Lynn Margulis was one of the most dedicated advocates of this more balanced view of the evolution of life. Of her several fine books, one she co-authored with her son Dorion Sagan, *Microcosmos*, gives perhaps the best account of life from this point of view, though some of the details have changed since it came out in 1992.
2. The best way to do this is to read the science writer Carl Zimmer's excellent *Microcosm*, which considers how life works through the lens of *E. coli*. It is usefully complemented by Maureen O'Malley's recent *Philosophy of Microbiology*, which also includes helpings of biology and history.
3. See McFall-Ngai et al (2013)

Chapter 3

1. Staley (1985)

2. See Singer (2013)

3. Salter (2014)

4. Cressey (2014)

5. Biologist Rob Dunn reckons media reports of the original germ-free animals led to our enthusiasm for eliminating germs in human environments. I think this is a complete misreading of the history, but he tells the story well. See Dunn (2011), chapter 5.

6. Quoted from Mazmanian (2009).

7. In a keynote at 'Exploring Human Host-Microbiome Interactions in Health and Disease', Hinxton, April 2014.

Chapter 4

1. See Human Microbiome Project Consortium (2012)

2. See Keeney et al (2014)

3. Hanage (2014)

4. Naik (2012)

5. See Fierer et al, (2010).

6. Sanford and Gallo (2013)

7. Lax (2014)

8. Hospodsky (2014)

9. See Nakatsuji (2013)

10. Kort (2014)

11. Eren (2014)

12. From a blog post, not the paper. See http://oligotyping.org/2014/06/25/oligotyping-analysis-of-the-human-oral-microbiome/

13. Ravel (2011)

14. Donia (2014)

15. Nelson (2012)

16. Price (2010)

17. This isn't a new idea. Thomas Luckey of the University of Missouri convened biennial symposia on 'intestinal microecology', starting in 1970. See, for example, Luckey (1972).

18. Dickson (2014)

19. Morrow (2010)

20. Dong (2011)

21. See Shaikh-Lesko (2014)

Chapter 5

1. Karlson (2014)
2. Moeller (2012)
3. Ding (2014)
4. Human Microbiome Project Consortium (2012)
5. Xu (2003)
6. McFall-Ngai (2013)
7. My story is based, among others, on Donahue (2011), Singh (2014), Yukihiro (2013, Bjarnadóttir (2006) and Ganapathy (2013).
8. Blaser's part of the *H. pylori* story forms the core of his recent book *Missing Microbes: How Killing Bacteria Creates Modern Plagues* (2014).
9. Hagymási (2014)
10. Saey (2014)

Chapter 6

1. Maynard (2012)
2. Aagaard (2014)
3. See Jeurink (2013)
4. Quoted in Pray (2012), p.80
5. A good recent review is Wopereis (2014). See also Koenig (2010)
6. Song (2014)
7. La Rosa (2014)
8. David (2014)
9. Claesson (2012)
10. Can (2014), Williams (2014)
11. Bosch (2012)
12. McCord (2013)
13. You can read the full argument in Richard Wrangham's *Catching Fire: How cooking made us human*. I'm convinced.
14. Tito (2012)
15. Adler (2012)
16. De Filippo (2010)
17. Schnorr (2014)
18. As reported in Yong (2014)
19. Moeller (2014)

Chapter 7

1. From Dr Fran Balkwill and Mic Rolph, *Cell Wars*, 1990 (ellipsis in the original). Another, slightly earlier popular guide to the immune system, from a small field, is titled *The Body At War* (Dwyer, 1988).

2. Scott Montgomery, in his brilliant *The Scientific Voice*, sums up all the metaphors that cluster round germs, disease, immunity and defence as 'biomilitarism', and traces their origins back to Pasteur, whose thoughts would have been shaped by France's defeat by Germany in the Franco-Prussian war of 1870–71. See Montgomery (1996).

3. Martin (1994), p.97

4. See Matzinger (2002), Pradeu (2012).

5. 'Paradigm shift' – Sommer (2013); 'Revolutionised our view of the immune system' – Belkaid (2014)

6. Thomas Prideau argues that bacteria have immunity as well, directed toward viruses. But this involves different cellular elements that recognise bits of nucleic acid and I think is too inclusive. Prideau (2012)

7. Estimate from Klein (1982)

8. Sonnenberg (2004)

9. McFall-Ngai (2007)

10. Lee (2010)

11. Eberl (2010)

12. Brandl (2008)

13. Bollinger (2007)

14. Everett (2004)

15. Lee (2010)

16. Belkaid, as above.

Chapter 8

1. Lee (2010), p.129

2. The story of Helminth worms and Crohn's disease is well told in Rob Dunn's *The Wildlife of our Bodies* (2011).

3. Mazmanian (2005, 2008)

4. Erickson (2012)

5. Gevers (2014)

6. Tang (2013), Loscalzo (2013)

7. Azad (2014)
8. Ley (2005)
9. Xiao (2013)
10. Cox (2013)
11. As reviewed in Dunne (2014)
12. Drawing here on Sachs' *Good Germs, Bad Germs* (2007)
13. Bach (2002)
14. A suggestion he is still promoting. See Rook (2012)
15. Rook (2013)

Chapter 9

1. Alison Bested and colleagues review all this in the first of three papers on the gut–brain story published in 2013, which I am drawing on here. Bested (2013).
2. For example in his book – Lyte (2010) – Chapter 1.
3. Lyte (2014)
4. House (2011)
5. Other possible bacterial influences on human appetite are canvassed in Norris (2013)
6. Bercik (2011)
7. All this and much more reviewed in Foster (2013)
8. Hsiao, Elaine (2013)
9. Cao (2013)
10. Tillisch (2013)
11. One small part of a complex series of studies by Ted Kaptchuk's group at Harvard that have been dissecting many aspects of placebo responses. See http://programinplacebostudies.org/. The more I think about it, the more this one becomes one of my favourite scientific studies. Kaptchuk (2010)

Chapter 10

1. Dennehy (2014)
2. As for the life of a bacterium (as referenced in Chapter 2) the best recent account of viruses is from the excellent Carl Zimmer, in 2011, from whom I take this comparison.
3. Ogilvie (2013)
4. Colson (2010)

5. Pride (2012)
6. Modi (2013)
7. Barr (2013)
8. Duerkop (2012)
9. Minot (2013)

Chapter 11

1. Eiseman (1958)
2. Youngster (2014)
3. See Smith (2014)
4. De Vrieze (2013)
5. Seekatz (2014)
6. Shanahan (2014)
7. Smith (2014)
8. $32.6 billion in 2014. See http://www.prnewswire.com/news-releases/marketsandmarkets-global-probiotics-market-worth-us326-billion-by-2014-81304537.html
9. Gibbons (2014).
10. Erdman (2014)
11. Kankainen (2009)
12. Ciorba (2012)
13. Allen (2013)
14. Petrof (2013)
15. Khanna (2014)
16. Callewaert (2014)
17. Sachs (2007), p204-205
18. You can check for signs of progress here: http://www.oragenics.com/?q=cavity-prevention
19. Sullivan (2011)
20. As plays out rather convincingly in Richard Powers' excellent novel *Orfeo* (2014).
21. The only details of this so far are in a press release: http://news.rice.edu/2014/05/12/no-bioengineered-gut-bacteria-no-glory-2/
22. Scott (2014)
23. An idea taken seriously by at least one student of the microbes in our artificial environments. See Upbin, 2013.
24. Maxman (2014)

25. Simons (2014)
26. Paxson (2008)
27. See http://www.design.philips.com/about/design/designportfolio/ design_futures/microbial_home.page
28. Richmond (2014)
29. Details here from the excellent book describing the whole project, *Synthetic Aesthetics*, Ginsberg (2014).
30. Sterling (2002), p.13.

Chapter 12

1. A distillation he expands at book length in Pollan (2006).
2. Some of these are taken from Thaiss (2014).
3. For an up-to-date review see Keeney (2014).
4. Karr (2012), Freddolino (2012)
5. See Noble (2006) for a provocative introduction to his approach.
6. Greenblum (2013). A comprehensive review which begins by conceding that 'The development of a predictive systems-level model of the microbiome represents a major leap forward and may remain out of reach for many years to come'.
7. Schluter (2012)
8. These images are usefully compiled in Nerlich (2009).
9. This is a rewording of a key passage in Richard Dawkins' *The Selfish Gene*, but taken from Denis Noble's account of a contrary view of genes' role in the organism. Noble (2006), p.12.
10. Alternative senses of the term I learn from O'Malley (2014).
11. O'Malley (2014) uses 'collaboration', too, and defines it: 'flexible symbiotic relationships that have both opt-in and opt-out possibilities, and which involve fluid functional rather than fixed taxonomic relationships.'
12. These points emerging from pondering an unpublished essay on the microbial superorganism idea by Jessica Houf.
13. Turney (2003)
14. Brodie (2007)
15. Joyce (2014)

Bibliography

Aagaard, Kjersti, et al (2014). The Placenta Harbors a Unique Microbiome. *Science Translational Medicine*, 6: 237

Abeles, Shiar and David Pride (2014). Molecular Bases and Role of Viruses in the Human Microbiome. *Journal of Molecular Biology*, 07.002

Adler, Christina, et al (2012). Sequencing ancient calcified dental plaque shows changes in oral microbiota with dietary shifts of the Neolithic and Industrial Revolutions. *Nature Genetics*, 45: 450–455

Alcock, Joe, et al (2014). Is eating behaviour manipulated by the gastrointestinal microbiota? Evolutionary pressures and mechanisms, *Bioessays*, 36: 201400071

Allen, Stephen (2013). *Lactobacilli* and *Bifidobacteria* in the prevention of antibiotic-associated diarrhoea and *Clostridium difficile* diarrhoea in older patients (PLACIDE): a randomised, double-blind, placebo-controlled, multicentre trial. *The Lancet*, 382: 1249–1257

Azad, M. (2014). Infant antibiotic exposure and the development of childhood overweight and central adiposity. *International Journal of Obesity*, doi:10.1038/ijo.2014.119

Azad, Meghan, et al (2013). Gut microbiota of healthy Canadian infants: profiles by mode of delivery and infant diet at 4 months. *Canadian Medical Association Journal*, 185: 385–394

Bach, J. (2002). The effect of infections on susceptibility to autoimmune and allergic diseases. *New England Journal of Medicine*, 347: 911–20

Bain, Robert, et al (2014). Global assessment of exposure to faecal contamination through drinking water based on a systematic review. *Tropical Medicine and International Health*. 19: 917–927

Bakalar, Nicholas (2003). *Where the Germs Are: A Scientific Safari*. John Wiley

Barr, Jeremy, et al (2013). Bacteriophage adhering to mucus provide a non-host-derived immunity. *Proceedings of the National Academy of Sciences*. 110: 10771–10776

Belkaid, Yasmine, and Timothy Ward (2014). Role of the Microbiota in Immunity and Inflammation. *Cell*, 157: 121–141

Bercik, P., et al (2011). The intestinal microbiota affect central levels of brain-derived neurotropic factor and behavior in mice. *Gastroenterology*, 141:599–609

Bested, Alison, et al (2013). Intestinal microbiota, probiotics and mental health: from Metchnikoff to modern advances. Part 1 – autointoxication revisited. *Gut Pathogens*, 5: 5

Bianconi, Eva, et al (2013). An estimation of the number of cells in the human body. *Annals of Human Biology*, 40: 463–471

Bjarnadóttir, T.K. et al (2006). Comprehensive repertoire and phylogenetic analysis of the G protein-coupled receptors in human and mouse. *Genomics* 88: 263–73

Blaser, Martin (2014). *Missing Microbes: How Killing Bacteria Creates Modern Plagues*. Oneworld

Bollinger, R.R., et al (2007). Biofilms in the large bowel suggest an apparent function of the human vermiform appendix. *Journal of Theoretical Biology*. 249: 826–831

Borre, Yuliya, et al (2014). Microbiota and neurodevelopmental windows: implications for brain disorders. *Trends in Molecular Medicine*, XX, 1–10

Bosch, Thomas (2012). What *Hydra* has to say about the role and origin of symbiotic interactions. *Biological Bulletin*, 223: 78–84

Bosch, Thomas (2013). Cnidarian–Microbe Interactions and the Origin of Innate Immunity in Metazoans. *Annual Review of Microbiology*, 67: 499–518

Brandl, K., et al (2008). Vancomycin-resistant enterococci exploit antibiotic-induced immune deficits. *Nature*, 455: 804–7

Brodie, Eoin, et al (2007). Urban aerosols harbor diverse and dynamic bacterial populations. *Proceedings of the National Academy of Sciences*, 104:299–304

Callewaert, Chris, et al (2014). Deodorants and antiperspirants affect the axillary bacterial community. *Archives of Dermatological Research*, 1007/s00403-014-1487-1

Can, Ismail, et al (2014). Distinctive thanatomicrobiome signatures found in the blood and internal organs of humans. *Journal of Microbiological Methods*, 106:1–7

Cao, Xinyi, et al (2013). Characteristics of the gastrointestinal microbiome in children with autism spectrum disorder: a systematic review. *Shanghai Archives of Psychiatry*. 25: 342–353

Ciorba, Matthew (2012). A Gastroenterologists's Guide to Probiotics. *Clinical Gastroenterology and Hepatology*, 10:960–968

Claesson, M., et al (2012). Gut microbiota composition correlates with diet and health in the elderly. *Nature*, 488: 178–84

Clemente, Jose, et al (2012). The Impact of the Gut Microbiota on Human Health: An Integrative View. *Cell* 148, March

Collins, Stephen (2012). The interplay between the intestinal microbiota and the brain. *Nature Reviews Microbiology*. 10: 735–742

Colson, P., et al (2010). Pepper mild mottle virus, a plant virus associated with specific immune responses, fever, abdominal pains, and pruritus in humans *PLoS One* 2010;5;e10041

Colson, P., et al (2013). Evidence of the megavirome in humans. *Journal of Clinical Virology*, 57: 191–200

Conrad, Peter (1999). A mirage of genes. *Sociology of Health and Illness*, 21, 228–241

Courage, Katherine (2014). Why Is Dark Chocolate Good for You? Thank Your Microbiome. ScientificAmerican.com. March 19

Cox, Laura, and Martin Blaser (2013). Pathways in Microbe-Induced Obesity, *Cell Metabolism*, 17: 883–894

Cressey, Dan (2014). Microbiome science threatened by contamination – Non-sterile sequencing returns rogue results in dilute microbiome samples. *Nature News*, 12 November. doi:10.1038/nature.2014.16327

Dalmasso, Marion, et al (2014). Exploiting gut bacteriophages for human health. *Trends in Microbiology*, 22: 399–405

Daston, Lorraine, and Elizabeth Lunbeck, eds (2011). *Histories of Scientific Observation*. Chicago University Press

David, Lawrence, et al (2014). Host lifestyle affects human microbiota on daily timescales. *Genome Biology*, 15/7/R89

De Filippo, C., et al (2010). Impact of diet in shaping gut microbiota revealed by a comparative study in children from Europe and rural Africa. *Proceedings of the National Academy of Sciences.* 107: 14691–14696

De Vrieze, Jop (2013). The Promise of Poop. *Science,* 341: 954–957

Dennehy, John (2014). What Ecologists Can Tell Virologists. *Annual Review of Microbiology,* 68: 117—35

Dickson, Robert, et al (2014). Towards an ecology of the lung: new conceptual models of pulmonary microbiology and pneumonia pathogenesis. *The Lancet Respiratory Medicine,* 2: 238–246

Ding, Tao and Patrick Schloss (2014). Dynamics and associations of microbial community types across the human body. *Nature,* 509: 357–360

Donahue, Dallas, et al (2011). The Microbiome and Butyrate Regulate Energy Metabolism and Autophagy in the Mammalian Colon. *Cell Metabolism,* 13: 517–5269

Dong, Q., et al (2011). Diversity of bacteria at healthy human conjunctiva. *Investigations in Ophthalmology and Vision Science,* 52: 5408–13

Donia, Mohamed, et al (2014). A Systematic Analysis of Biosynthetic Gene Clusters in the Human Microbiome Reveals a Common Family of Antibiotics. *Cell,* 158: 1402–1414

Duerkop, Breck, et al (2012). A composite bacteriophage alters colonization by an intestinal commensal bacterium. *Proceedings of the National Academy of Sciences,* 109, 17621–17626

Dunne, J., et al (2014). The intestinal microbiome in type 1 diabetes. *Clinical & Experimental Immunology,* 177: 30–37

Dunn, Rob (2011). *The Wild Life of Our Bodies – Predators, Parasites, and Partners That Shape Who We Are Today.* HarperCollins

Dutilh, Bas, et al (2014). A highly abundant bacteriophage discovered in the unknown sequences of human faecal metagenomes. *Nature Communications,* 5: 4498

Eberl, G. (2010). A new view of immunity: homeostasis of the superorganism. *Mucosal Immunology,* 3: 450–460

Eiseman, B., et al (1958). Fecal enema as an adjunct in the treatment of pseudomembranous colitis. *Surgery,* 44:854–859

Erdman, S., and T. Poutahadis (2014). Probiotic 'glow of health': it's more than skin deep. *Beneficial Microbes*, 5: 109–119

Eren, A. Murat, et al (2014). Oligotyping analysis of the human oral microbiome. *Proceedings of the National Academy of Sciences*, June 25, E2875–2884

Erickson, Alison, et al (2012). Integrated Metagenomics/ Metaproteomics Reveals Human Host–Microbiota Signatures of Crohn's Disease. *PloS One*. 7: e49138

Everett, M.L., et al (2004). Immune Exclusion and Immune Inclusion: a New Model of Host–Bacterial Interactions in the Gut. *Clinical and Applied Immunology Reviews*. 5: 321–332

Fierer, Noah, et al (2010). Forensic identification using skin bacterial communities. *Proceedings of the National Academy of Science*, 107: 6477–6481

Foster, Jane, and Karen-Anne Neufeld (2013). Gut–brain axis: how the microbiome influences anxiety and depression. *Trends in Neurosciences*, 36: 305–312

Fowler, R. (1986). Howard Hughes: A psychological autopsy. *Psychology Today*, May, 22–33

Freddolino, Peter, and Saeed Tavazoie (2012). The Dawn of Virtual Cell Biology. *Cell*, 150: 248–251

Funkhouser, Lisa, and Seth Bordenstein (2013). Mom Knows Best: The Universaility of Maternal Microbial Transmission. *PloS Biology*. 11: e1001631

Furusawa, Yukihiro, et al (2013). Commensal microbe-derived butyrate induces the differentiation of colonic regulatory T cells. *Nature*, 504: 446–450

Ganapathy, Vadivel, et al (2013). Transporters for short-chain fatty acids as the molecular link between colonic bacteria and the host. *Current Opinion in Pharmacology*, 13: 869–874

Gest, Howard (2004). The Discovery of Microorganisms by Robert Hooke and Antoni van Leeuwenhoek, Fellows of the Royal Society. *Notes and Records of the Royal Society of London*, 58: 187–201

Gevers, Dirk, et al (2014). The Treatment-Naïve Microbiome in New-Onset Crohn's Disease. *Cell Host & Microbe*. 15: 382–392

Gibbons, Ann (2014). The Evolution of Diet. *National Geographic*, August

Ginsberg, Alexandra, et al (2014). *Synthetic Aesthetics: Investigating Synthetic Biology's Designs on Nature*. MIT Press

Goldenberg, J., et al (2013). Probiotics for the prevention of *Clostridium difficile* associated diarrhea in adults and children. *Cochrane Database Systematic Reviews*. 5: CD006095

Greenblum, Sharon, et al (2013). Towards a predictive systems-level model of the human microbiome: progress, challenges, and opportunities. *Current Opinion in Biotechnology*, 24:810–820

Hagymási, Kristina, et al (2014). *Helicobacter pylori* infection: New pathogenetic and clinical aspects. *World Journal of Gastroenterology*. 20: 6386–6399

Hanage, William (2014). Microbiome science needs a healthy dose of scepticism. *Nature*, 512: 247–8

Hickey, Roxana, et al (2012). Understanding vaginal microbiome complexity from an ecological perspective. *Translational Research*, 160: 267–282

Hornick, R., et al (1970). Typhoid Fever: Pathogenesis and Immunologic Control. *New England Journal of Medicine*, 283: 739–746

Hospodsky, D., et al (2014). Hand bacterial communities vary across two different human populations. *Microbiology*, 160:1144–1152

House P.K., et al (2011). Predator cat odors activate sexual arousal pathways in brains of *Toxoplasma gondii* infected rats. *PLoS ONE* 6: e23277

Hsiao, Elaine, et al (2013). Microbiota Modulate Behavioral and Physiological Abnormalities Associated with Neurodevelopmental Disorders. *Cell*, 155: 1451–1463

Human Microbiome Project Consortium (2012). A framework for human microbiome research. *Nature*, 486: 215–219

Human Microbiome Project Consortium (2012). Structure, function and diversity of the healthy human microbiome. *Nature*, 486: 207–214

Jackson, Mark (2006). *Allergy: The History of a Modern Malady*. Reaktion Books

Jeurink, P., et al (2013). Human milk: a source of more life than we imagine. *Beneficial Microbes*, 4: 17–30

Joyce, Christopher (2014). Soil Doctors Hit Pay Dirt In Manhattan's Central Park. http://www.npr.org/2014/10/02/353066730/

Kankainen, M., et al (2009). Comparative genomic analysis of *Lactobacillus rhamnosus GG* reveals pili containing a human-mucus binding protein. *Proceedings of the National Academy of Sciences*. 106:17193–17198

Kaptchuk, Ted, et al (2010). Placebos without Deception: A Randomised Controlled Trial in Irritable Bowel Syndrome. *PloS One*. 0015591

Karlsson, F.H., et al (2014). Metagenomic Data Utilization and Analysis (MEDUSA) and Construction of a Global Gut Microbial Gene Catalogue. *PLoS Comput Biol* 10: e1003706

Karr, Jonathan, et al (2012). A Whole-Cell Computational Model Predicts Phenotype from Genotype. *Cell*, 150:398-401

Keeney, Kristie, et al (2014). Effects of Antibiotics on Human Microbiota and Subsequent Disease. *Annual Review of Microbiology*. 68: 217–35

Khanna, Sahil, et al (2014). Clinical Evaluation of SER-90, a Rationally Designed, Oral Microbiome-Based Therapeutic for the Treatment of Recurrent *Clostridium difficile*. Orally delivered at American Gastroenterological Association conference, August. See http://www.gastro.org/news/articles/2014/08/14/new-frontiers-of-fecal-microbiota-transplantation

Klein, Jan (1982). *Immunology: The Science of Self–Nonself Discrimination*. John Wiley

Koenig, Amy, et al (2010). Succession of microbial consortia in the developing infant gut microbiome. *Proceedings of the National Academy of Sciences*, 1000081107

Kort, Remco, et al (2014). Shaping the oral microbiota through intimate kissing. *Microbiome*, 2:41

Kotula, Jonathan, et al (2014). Programmable bacteria detect and record an environmental signal in the mammalian gut. *Proceedings of the National Academy of Sciences*, 111: 4838–4843

Kroes, I. et al (1999). Bacterial diversity within the human subgingival crevice, *Proceedings of the National Academy of Sciences* 96: 14547–52

La Rosa, Patricio, et al (2014). Patterned progression of bacterial populations in the premature infant gut. *Proceedings of the National Academy of Sciences*, 111:12522–12527

Lax, S., et al (2014). Longitudinal analysis of microbial interaction between humans and the indoor environment. *Science*, 345: 1048–1052

Lecuit, Marc and Marc Eloit (2013). The human virome: new tools and new concepts. *Trends in Microbiology*, 21: 510–515

Lederberg, Joshua (2000). Infectious History. *Science*, 288: 287–293

Lee, Stewart (2010). *How I Escaped my Certain Fate – The Life and Deaths of a Stand-Up Comedian*. Faber

Lee, Yun, and Sarkis Mazmanian (2010). Has the Microbiota Played a Critical Role in the Evolution of the Adaptive Immune System? *Science* 330: 1195568

Ley, R., et al (2005). Obesity alters gut microbial ecology. *Proceedings of the National Academy of Sciences*, 102: 11070–11075

Li, Juanha, et al (2014). An Integrated Catalogue of reference genes in the human gut microbiome, *Nature Biotechnology*, published online 6 July 2014

Loscalzo, Joseph (2013). Gut Microbiota, the Genome, and Diet in Atherogenesis. *New England Journal of Medicine*, 368:17

Luckey, T. (1972). Introduction to intestinal microecology. *American Journal of Clinical Nutrition*. 25: 1292–1294

Lyte, Mark (2014). Microbial endocrinology: Host–microbiota neuroendocrine interactions. *Gut Microbes*, 5: 28682

Lyte, Mark, and Primrose Freestone, eds (2010). *Microbial Endocrinology: Interkingdom Signalling in Infectious Disease and Health*. Springer

Martin, Emily (1994). *Flexible Bodies: The role of immunity in American culture from the days of polio to the age of AIDS*. Beacon Press.

Matzinger, P. (2002). The danger model: a renewed sense of self. *Science*, 296: 301–305

Maxman, A. (2013). Designing for microscopic life in the great indoors (Interview with Jessica Green). *New Scientist*, 20 July

Maynard, Craig, et al (2012). Reciprocal interactions of the intestinal microbiota and immune system. *Nature*, 489: 231–241

Mazmanian, S., et al (2005). An immunomodulatory molecule of symbiotic bacteria directs maturation of the host immune system. *Cell*, 122: 107–118

Mazmanian, Sarkis, et al (2008). A microbial symbiosis factor prevents intestinal inflammatory disease. *Nature*, 453: 620–625

Mazmanian, Sarkis (2009). Microbial Health Factor. *The Scientist*, August 1

McCord, Aleia et al (2013). Faecal Microbiomes of Non-Human Primates in Western Uganda Reveal Species-Specific Communities Largely Resistant to Habitat Perturbation. *American Journal of Primatology*, 22238

McFall-Ngai, Margaret (2007). Care for the community. *Nature*, 445: 153

McFall-Ngai, Margaret, et al (2013). Animals in a bacterial world, a new imperative for the life sciences. *Proceedings of the National Academy of Sciences*, 110, 3229–3236

Minot, S., et al (2013). Rapid Evolution of the Human Gut Virome. *Proceedings of the National Academy of Sciences*, 110: 12450–12455

Modi, Sheetal (2013). Antibiotic treatment expands the resistance reservoir and ecological network of the phage metagenome. *Nature*, 499: 219–223

Moeller, Andrew, et al (2012). Chimpanzees and humans harbour compositionally similar gut enterotypes. *Nature Communications*, 3: 1179

Moeller, Andrew, et al (2014). Rapid changes in the gut microbiome during human evolution. *Proceedings of the National Academy of Sciences*, Nov 3. doi: 10.1073/pnas.1419136111

Montgomery, Scott (1996). *The Scientific Voice*. Guilford Press

Morrow, L., et al (2010). Probiotic prophylaxis of ventilator-associated pneumonia: a blinded, randomized, controlled trial. *American Journal of Respiratory & Critical Care Medicine*. 182: 1058–1064

Naik, Shruti, et al (2012). Compartmentalized Control of Skin Immunity by Resident Commensals. *Science*, 337: 1115–1119

Nakatsuji, Teruaki, et al (2013). The microbiome extends to subepidermal compartments of normal skin. *Nature Communications*, 4. Feb 5

Nelson, David, et al (2012) Bacterial Communities of the Coronal Sulcus and Distal Urethra of Adolescent Males. *PloS One*, 7: e36298

Nerlich, Brigitte, and Lina Hellsten (2009). Beyond the human genome: microbes, metaphors and what it means to be human in an interconnected post-genomic world. *New Genetics and Society*, 28: 19–36

Noble, Denis (2006). *The Music of Life: Biology Beyond Genes*. Oxford University Press

Norris, Vic, et al (2013). Hypothesis: Bacteria Control Host Appetites. *Journal of Bacteriology*, 195:411

O'Malley, Maureen (2014). *Philosophy of Microbiology*. Oxford University Press

Ogilvie, Lesley, et al (2013). Genome signature-based dissection of human gut metagenomes to extract subliminal viral sequences. *Nature Communications*, 3420

Okada, H., et al (2010). The 'hygiene hypothesis' for autoimmune and allergic diseases: an update. *Clinical & Experimental Immunology*, 160: 1–9

Paxson, Heather (2008). Post-Pasteurian Cultures: The Microbiopolitics of Raw-Milk Cheese in the United States. *Cultural Anthropology*, 23: 15–47

Petrof, E., et al (2013). Microbial ecosystems therapeutics: a new paradigm in medicine? *Beneficial Microbes* 4: 53–65

Petrof, E., et al (2013). Stool substitute transplant therapy for the eradication of *C. difficile* infection: RePOOPulating the gut. *Microbiome* 1:3

Pollan, Michael (2006). *The Omnivore's Dilemma: A Natural History of Four Meals*. Penguin

Pradeu, Thomas (2012). *The Limits of the Self: Immunology and Biological Identity*. Oxford University Press

Pradeu, Thomas, and Edwin Cooper (2012). The danger theory: 20 years later. *Frontiers in Immunology*, 3: 287

Pray, Leslie, et al (2012). *The Human Microbiome, Diet and Health: Workshop Summary*. Institute of Medicine, US

Price, Lance (2010). The Effects of Circumcision on the Penis Microbiome. *PloS One*, 5: e8422

Pride, David, et al (2012). Evidence of a robust resident bacteriophage population revealed through analysis of the human salivary genome. *ISME Journal*, 6: 915–926

Ravel, J., et al (2011). Vaginal microbiome of reproductive-age women. *Proceedings of the National Academy of Sciences.* 108 Suppl 1:4680–7

Relman, David (2012). The human microbiome: ecosystem resilience and health. *Nutrition Reviews,* 70: S2–S9

Relman, David, et al (2009). *Microbial Evolution and Co-Adaptation – A Tribute to the Life and Scientific Legacies of Joshua Lederberg.* National Academy of Sciences, US

Richmond, Ben (2014). Cola-flavoured Genitals, and Other Potential Uses for Microbiome Hacking. http://motherboard.vice.com/ September 11

Ridaura, V., et al (2013). Gut microbiota from twins discordant for obesity modulate metabolism in mice. *Science,* 341: 1241214

Roach, Mary (2013). *Gulp: Adventures in the Alimentary Canal.* Oneworld

Rook, Graham (2012). A Darwinian view of the Hygiene or 'Old Friends' hypothesis. *Microbe,* 7: 173–180

Rook, Graham (2013). Regulation of the immune system by biodiversity from the natural environment: An ecosystem service essential to health. *Proceedings of the National Academy of Sciences,* 110: 18360–18367

Rosebury, Theodor (1969). *Life on Man.* Secker and Warburg

Sachs, Jessica Snyder (2007). *Good Germs, Bad Germs: Health and Survival in a Bacterial World.* Hill and Wang

Saey, Tina (2014). Here's the poop on getting your gut microbiome analysed. www.sciencenews.org June 17

Salter, Susannah, et al (2014). Reagent and laboratory contamination can critically impact sequence-based microbiome analysis. *BMC Biology,* 12: 87

Sanford, James, and Richard Gallo (2013). Functions of the skin microbiota in health and disease. *Seminars in Immunology,* 25: 370–377

Scheuring, Istvan, and Douglas Wu (2012). How to assemble a beneficial microbiome in three easy steps. *Ecology Letters,* 15: 1300–1307

Schluter, Jonas, and Kevin Foster (2012). The Evolution of Mutualism in Gut Microbiota via Epithelial Selection. *PLOS Biology,* 10: e1001424

Schnorr, Stephanie, et al (2014). Gut microbiome of the Hadza hunter-gatherers. *Nature Communications,* 5: 3654

Scott, Julia (2014). My No-Soap, No-Shampoo, Bacteria-Rich Hygiene Experiment. *New York Times Magazine*, May 22

Seekatz, Anna, et al (2014). Recovery of the Gut Microbiome following Fecal Microbiota Transplantation. *mBio*, 5: e00893

Shaikh-Lesko, Rina (2014). Visualising the ocular microbiome. *The Scientist*, May 12

Shanahan, Fergus, and Eamonn Quigley, (2014). Modification of the gut microbiome to maintain health or treat disease. *Gastroenterology*, 146: 1554–1563

Simons, Jake (2014). *Biodesign: Why the future of our cities is soft and hairy.* http://edition.cnn.com/2014/08/27/tech/innovation/biodesign-why-the-future-of-our-cities-is-soft-and-hairy/

Singer, Emily (2013).Our bodies, our data. *Quanta Magazine* online. October. http://www.simonsfoundation.org/quanta/20131007-our-bodies-our-data/

Singh, Nagendra, et al (2014). Activation of Gpr109a, Receptor for Niacin and the Commensal Metabolite Butyrate, Suppresses Colonic Inflammation and Carcinogenesis. *Immunity*, 40: 128–139

Smith, Mark, et al (2014). How to regulate faecal transplants. *Nature*, 506: 290–91

Smith, Peter (2014). Is your body mostly microbes? Actually, we have no idea. *Boston Globe*, September 14

Smith, Peter (2014). A New Kind of Transplant Bank. *New York Times*, Feb 17

Sommer, Felix, and Fredrik Bäckhed (2013). The gut microbiota – masters of host development and physiology. *Nature Reviews Microbiology*, 11: 227–238

Song, Se Jin, et al (2013). How delivery mode and feeding can shape the bacterial community in the infant gut. *Canadian Medical Association Journal*, 185: 373–374

Song, Se Jin, et al (2013). Cohabiting family members share microbiota with one another and with their dogs. *eLife*, 2: e00458

Sonnenberg, Justin, et al (2004). Getting a grip on things: how do communities of bacterial symbionts become established in our intestine? *Nature Immunology*, 5:569–573

Staley, James and Allen Konopka (1985). Measurement of in situ activities of nonphotosynthetic microorganisms in aquatic and terrestrial habitats. *Annual Review of Microbiology*, 39: 321–46

Sterling, Bruce (2002). *Tomorrow Now: Envisioning the Next Fifty Years*. Random House

Sullivan, et al (2011). Clinical Efficacy of a Specifically Targeted Antimicrobial Peptide Mouth Rinse: Targeted Elimination of *Streptococcus mutans* and Prevention of Demineralization. *Caries Research*. 45: 415–428

Tang, W., et al (2013). Intestinal Microbial Metabolism of Phosphatidylcholine and Cardiovascular Risk. *New England Journal of Medicine*, 368: 1575–1584

Thaiss, Christoph, et al (2014). *Exploring New Horizons in Microbiome Research*. Cell, Host and Microbe. 15: 662–667

Tillisch, Kirsten, et al (2013). Consumption of Fermented Milk Product With Probiotic Modulates Brain Activity. *Gastroenterology*, 144: 1394–1401

Tito, Raul (2012). Insights from Characterizing Extinct Human Gut Microbiomes. *PloS One*. 7: e51146

Tomes, Nancy (1998). *The Gospel of Germs. Men, Women and the Microbe in American Life*. Harvard University Press

Turney, J. (2003). *Lovelock and Gaia: Signs of Life*. Icon Books.

Upbin, Bruce (2013). Yogurt is probiotic, why not your steering wheel? Forbes.com 3/01/2013

Van Nood, Els, et al (2013). Duodenal Infusion of Donor Feces for Recurrent *Clostridium difficile*. *New England Journal of Medicine*, 368: 407–415

Waller, John (2002). *The Discovery of the Germ*. Icon Books

Wilcox, Christie (2013). Fake Feces To Treat Deadly Disease: Scientists Find They Can Make Sh*t Up. scientificamerican.com, January 10

Williams, Anna (2014). Your death microbiome could catch your killer. *New Scientist*, 27 August

Wolf, M. (2014). Is there really such a thing as 'one health'? Thinking about a more than human world from the perspective of cultural anthropology, *Social Science & Medicine* 2014.06.018

Wopereis, Harm, et al (2014). The first thousand days – intestinal microbiology of early life: establishing a symbiosis. *Pediatric Allergy and Immunology*, March

Wrangham, Richard (2009). *Catching Fire: How Cooking Made Us Human*. Basic Books

Wylie, Kristine, et al (2012). Emerging view of the human virome. *Translational Research*, 160: 283–290

Xiao, Shuiming, et al (2013). A gut microbiota-targeted dietary intervention for amelioration of chronic inflammation underlying metabolic syndrome. *FEMS Microbial Ecology*, 87: 357–367

Xu, Jian, and Jeffrey Gordon (2003). Honor thy symbionts. *Proceedings of the National Academy of Sciences*, 100: 10452–10459

Yatsuneko, Tanya, et al (2012). Human gut microbiome viewed across age and geography. *Nature*, 486: 222–228

Yong, Ed (2014). Searching for a 'Healthy' Microbiome. http://www.pbs.org/wgbh/nova/next/body/microbiome-diversity/ 29 Jan 2014.

Youngster, Ilan, et al (2014). Oral, Capsulized, Frozen Fecal Microbiota Transplantation for Relapsing *Clostridium difficile* Infection. *Journal of the American Medical Association*, Oct 11, doi:10.1001/jama.2014.13875

Zhao, L. (2013). The gut microbiota and obesity: from correlation to causality. *Nature Reviews Microbiology*. 11: 639–647

Zimmer, Carl (2008). *Microcosm: E. coli and the new science of life*. Heinemann.

Zimmer, Carl (2011). *A Planet of Viruses*. University of Chicago Press.

Zipursky, Jonathan, et al (2012). Patient Attitudes Toward the Use of Fecal Microbiota Transplantation in the Treatment of Recurrent *Clostridium difficile* Infection. *Clinical Infectious Diseases*, 55: 1652–1658

Further reading

Research on the human microbiome, and those of other creatures, is expanding fast. This book gives some useful pointers to what to look out for in future results, and most of the open questions I mention will be clarified gradually rather than through sudden breakthroughs.

Still, a topic like this will see important studies that I haven't covered. Much of the work appears in 'open access' journals nowadays, which is great, but most of us will need translations as well. The literature is scattered, too. Microbiome papers turn up in obvious places (there are journals now with the word in their titles), but also in general life sciences journals, and titles in particular medical specialities, in immunology, cell biology, ecology … You can see which ones publish which kind of thing from the notes and bibliography here, and most publish review articles regularly because of the interest in the field just now. The news pages of the professional journals *Science* and *Nature* – not themselves open access but some of their writing is – will help, as will the more layperson-friendly *Scientific American*.

The best interpreters of new research are easy to follow on the web: from the UK Ed Yong's blog 'Not Exactly Rocket Science', and from the US Carl Zimmer's 'Loom' regularly offer highly thoughtful and readable microbiome updates. Both are in the National Geographic blog stable: http://phenomena. nationalgeographic.com/blog/not-exactly-rocket-science/ and http://phenomena.nationalgeographic.com/blog/the-loom/

Elsewehere, researcher Jonathan Eisen posts regularly at 'The Tree of Life' (http://phylogenomics.blogspot.co.uk/), and helps keep us writers honest with his 'overselling the microbiome' award.

All these good folk are also on twitter, as is Elisabeth Bik (@MicrobiomDigest), who will keep you abreast of newly published papers and commentary on microbiome results on a daily basis at http://www.microbiomedigest.com/

Happy reading.

Index